Tissue Culture Methods
for Plant Pathologists

Frontispiece   Georges Morel (1916-1973), whose great foresight
and imagination led him to pioneer the application of tissue culture
methods in plant pathology. This photograph, taken in the laboratory
in 1964, was supplied by Jean-Pierre Bourgin, *Institut National de
la Recherche Agronomique, Laboratoire de Biologie Cellulaire, CNRA,
Versailles, France.*

Tissue Culture Methods for Plant Pathologists

Edited for the

Federation of British Plant Pathologists

by

D. S. INGRAM
Botany School, Downing Street,
Cambridge CB2 3EA, UK

and

J. P. HELGESON
USDA, SEA, AR Plant Disease Resistance Research Unit
Department of Plant Pathology,
University of Wisconsin,
Madison, WI 53706, USA

BLACKWELL SCIENTIFIC PUBLICATIONS
OXFORD LONDON EDINBURGH
BOSTON MELBOURNE

(C) 1980 Blackwell Scientific Publications
Editorial offices:
Osney Mead, Oxford OX2 OEL
8 John Street, London WC1N 2ES
9 Forrest Road, Edinburgh EH1 2QH
52 Beacon Street, Boston, Massachusetts 02108, USA
214 Berkeley Street, Carlton, Victoria 3053, Australia

First published 1980

British Library
Cataloguing in Publication Data

Tissue culture methods for plant pathologists.
    1. Plant physiology - Research
    I. Ingram, David Stanley II. Helgeson, J P
    III. Federation of British Plant Pathologists
    581.072'4   QK725

    ISBN 0-632-00715-X

Distributed in the U.S.A. and Canada
by Halsted Press, a division of
John Wiley & Sons Inc, New York

Printed in Great Britain by
Galliard (Printers) Ltd, Great Yarmouth
and bound by
Mansell (Bookbinders) Ltd, Witham, Essex

The Federation of British Plant Pathologists is part of the Association
of Applied Biologists and the British Mycological Society

Contents

Section 1
Introduction

Section 2
Basic tissue culture methods

Section 3
Methods for use with viruses

A. Infection of protoplasts

B. Virus elimination

Section 4
Methods for use with fungi,
bacteria and nematodes

Preface

In December 1979 the Federation of British Plant Pathologists organized
a conference at Leeds University on Methods in Plant Pathology. One
session of the conference, convened for the Federation by D.S. Ingram
in consultation with F.J.W. England of the Plant Breeding Group of the
Association of Applied Biologists, was largely devoted to demonstrations
dealing with tissue culture methods for plant pathologists. This book
is based on those demonstrations.

During the past ten or fifteen years there has been considerable
progress in the development of plant tissue culture methods, and their
use in plant pathology has increased significantly. Today they are
being employed successfully for such varied purposes as studying host-
parasite interactions, growing obligate biotrophs in culture, freeing
plants of pathogens, preserving, propagating and transporting valuable
germplasm and breeding new disease-resistant cultivars. The demonstration topics, and hence the papers in the book, reflect this diversity,
although they do not represent a comprehensive survey.

We believe that many more plant pathologists would like to use
tissue culture techniques in their research, but hesitate to do so
because of lack of information. Several good books dealing with plant
tissue culture already exist, but none of these have been written with
the plant pathologist in mind. We hope that the present book will
partly remedy this situation by introducing research workers, especially
graduate students and newcomers to the field, to a wide range of
established and recently developed methods and applications which can
either be used directly in research programmes or modified to suit
particular needs. In addition, we hope that much of the material will
be of use to University and College teachers as a basis for class
experiments and projects.

The first major section of the book deals with basic tissue culture
methods. To make it more complete two papers have been included which
are not based on demonstrations presented at Leeds. These deal with
root organ culture and callus and cell culture. Certain other methods
not referred to in the first section, such as meristem-tip and embryo
culture, are covered later in the book.

Further sections deal with applications of the basic methods in
various aspects of plant pathology and in breeding disease-resistant
plants. Some of the papers are presented as detailed step-by-step
instructions, others as more wide-ranging discussions of techniques and
objectives. The introductory paper by Ingram and the paper on plant

xii

breeding by Day are based on lectures given at Leeds. The concluding
discussion by Helgeson was especially written for the book.

   On behalf of the Federation we thank the authors for their patient
co-operation in making their papers conform to the pattern of the book.
We also thank T.F. Preece, D.J. Royle and C.J. Rawlinson, who all
played key roles in the organization of the conference.

   We are grateful to P.R. Scott for advice during the preparation of
the book, and also to A. Ingram who was responsible for the production
editing, J.E. Brown and K.S. Jacklin who prepared the camera-ready
typescript and M.J. Howe who compiled the index.

                         D.S. INGRAM
                         J.P. HELGESON

                         Federation of British Plant Pathologists

Section 1
Introduction

Tissue culture methods in plant pathology

D.S. INGRAM

Botany School, Downing Street,
Cambridge CB2 3EA

INTRODUCTION

The term tissue culture is used here in its widest sense to describe
the growth of isolated plant organs (roots, shoot tips, embryos etc.),
anthers and pollen (to produce haploids), callus tissues, cells and
protoplasts in culture on defined and semi-defined media, in the absence
of contaminant micro-organisms. A basic introduction to the practical
aspects of the procedures involved is included in this volume (Butcher;
Hanke; Helgeson; Henshaw et al.; Hussey; Sunderland; Thomas et al.;
Withers). Further information may be obtained from the following books:
Gautheret (1959); White (1963); Willmer (1966); Kruse & Patterson (1973);
Kasha (1974); Street (1974, 1977); Gamborg & Wetter (1975); Thomas &
Davey (1975); Butcher & Ingram (1976); Dudits et al. (1976); Peberdy et
al. (1976); Barz et al. (1977); Reinert & Bajaj (1977); Thorpe (1978);
Durbin (1979); Pierik (1979); Davies & Hopwood (1980).

Plant tissue culture techniques were originally developed for the
study of fundamental problems of nutrition and morphogenesis, and have
led to many important advances in these fields. More recently, as
techniques and media have improved, it has become increasingly clear
that tissue cultures provide simplified experimental systems for use in
many other branches of plant science, including plant pathology. Two
important limitations must be recognized, however. First, many of the
techniques available have been perfected for only a relatively small
number of plant species, notably tobacco, and much research is still
required to extend the lists of species involved. Second, cultured
cells and tissues, particularly callus and suspension cultures, may be
physiologically and genetically different from whole plants. Providing
that these limitations are firmly borne in mind when planning and
interpreting the results of experiments, and providing that culture
systems are properly defined, pitfalls may be avoided.

In addition to their use as experimental tools tissue cultures have
other, more practical, applications in plant pathology, especially in
the production of pathogen-free clones, the long-term storage of
germplasm and the generation of novel disease-resistant lines of crop

plants.

In this chapter the contribution of tissue culture technology to plant pathology is briefly reviewed, and some suggestions for further research are adumbrated. The chapters which follow provide a practical introduction to the procedures upon which such research may be based.

APPLICATIONS IN VIROLOGY

Protoplast culture

This subject has been reviewed recently by Zaitlin & Beachy (1974), Takebe (1975, 1977) and Wood *et al.* (1979).

Investigation of the molecular events involved in the penetration and replication of plant viruses was, in the past, hampered by the presence of the cell wall of the host. Thus the only method of inoculation available was to rub the surface of a leaf with a viral suspension containing an abrasive. This technique is inefficient, cannot be quantified and gives inconsistent results. The development of techniques for the inoculation of cultured protoplasts with virus or virus-RNA in the presence of polycations and other substances (Barker; Cassells & Cocker; Oldfield & Coutts; Watts *et al.*; Wood *et al.*; this volume) allowed the barrier of the cell wall to be circumvented. This created an experimental system with the following advantages: infection is synchronous, leading to one-step replication; a high proportion of cells become infected and there is a high efficiency of infection; protoplasts may be pipetted, so infected material can be handled quantitatively; experimental conditions can be precisely controlled; virus related molecules and structures may be isolated with ease.

There are also some limitations to the use of protoplasts in virus research: protoplasts are often difficult to handle and work with; there may be problems in extending the results of experiments made with protoplasts to the intact plant, because of the physiological and structural differences between the two systems; and although procedures for the infection and replication of tobacco mosaic virus (TMV) in tobacco protoplasts are well defined (Takebe, 1977) few other interactions have been studied extensively. There is an urgent need for research to extend the list of viruses so far grown in protoplast culture, especially to include more large viruses and viruses with divided genomes.

Despite the problems, protoplast culture has already led to many notable advances in virus research, especially in the field of replication of TMV. There is every reason to believe that this trend will continue. The following are among the topics currently under investigation (Takebe, 1977; Wood *et al.*, 1979).
  1. Infection.
  2. Replication and integration of the viral genome with that of the host.

3. Virus-virus interactions such as cross protection and related phenomena (Watts *et al.*; Barker; this volume).
4. Multiparticulate viruses.
5. Resistance and specificity (Wood *et al.*, this volume).
6. Symptomology, such as changes in organelles and the biogenesis of inclusion bodies.

Production of virus-free plants

Aspects of this subject have been reviewed by Hollings (1965), Murashige (1974), Quak (1977) and Walkey (1978).

Most crop plants which are propagated vegetatively contain viruses which affect performance or depress yield. Elimination of such viruses from many species may be effected by heat treatment, although some have resisted all attempts at eradication by this means. The most successful alternative so far devised is that of meristem-tip culture (Cassells *et al.*; Dale *et al.*; Walkey; this volume). Excised meristem-tips grow on to produce whole plants which may be free both of viruses and of other pathogens. After careful indexing such plants may be bulked-up and released to growers as stock. Improved results are often obtained when meristem-tips are taken from heat-treated plants or when antiviral chemicals are incorporated into culture media.

Meristem-tip culture is now possible with a wide range of ornamental and crop species (Walkey, this volume), including potato (Henshaw *et al.*, this volume) and forage legumes and grasses (Dale *et al.*, this volume). Research is still required to extend this list, however, while the following aspects of meristem-tip technology also require further investigation.
1. The physiological basis of virus elimination in culture.
2. The development and mode of action of anti-viral chemicals (Cassells & Long, this volume).
3. Procedures for *in vitro* replication of meristem-tips to allow rapid clonal multiplication of virus-free plants (Hussey, this volume).
4. Methods for the long-term storage of virus-free or valuable germplasm, such as cryopreservation and low growth (Henshaw *et al.*; Withers; this volume).
5. Alternative culture methods for the production of pathogen-free plants (Cassells *et al.*; Hussey; this volume).

APPLICATIONS IN THE STUDY OF FUNGI, BACTERIA AND NEMATODES

General reviews of this subject have been written by Ingram (1976, 1977). Applications of tissue culture techniques in studies of tumour physiology have been reviewed by Butcher (1977) and in studies of nematodes by Zuckerman (1971).

Obligate parasites

Plant tissue cultures have an obvious advantage for studies of that group of fungi and nematodes which cannot be grown easily in the laboratory on artificial media. This fact was originally recognized by Morel, who in the early 1940s infected callus from vine with zoospores of the downy mildew fungus *Plasmopara viticola*. Since then a number of other obligate biotrophs have been established in dual culture with their hosts including, for example, several downy mildews (Peronosporaceae; Ingram, this volume), some rusts (Uredinales; Ingram, 1977), especially those causing galls, and the intracellular biotroph *Plasmodiophora brassicae* (Buczacki, this volume). Unfortunately the powdery mildews (Erysiphaceae; Webb & Gay, this volume) have so far proved particularly difficult to deal with, and more research is needed to determine the reasons for this. Methods for establishing dual cultures of a wide range of parasitic nematodes and their hosts have also been devised (Jones, this volume).

With the exception of the establishment of nematode banks and some attempts to produce axenic cultures of various fungi, dual cultures have not been widely used as experimental tools. This is unfortunate, for they have several potential uses in areas such as the following.
1. The establishment of axenic cultures.
2. The maintaining and transport of supplies of aseptic inoculum for experimental use.
3. The study of sexual and asexual reproduction.
4. The study of the metabolic aspects of host-parasite interaction (see below).

Host-parasite interaction

Tissue cultures offer many advantages for the study of host-parasite interaction in a wide range of plant diseases caused by parasitic fungi, bacteria and nematodes. They offer the attractive possibility of a simplified experimental system where the physical and chemical environments are precisely controlled, large numbers of host cells may be exposed to a parasite without excessive tissue injury, cell numbers and inoculum density may be precisely controlled and the precursors and products of interaction may be added to, or removed from, the culture system with ease.

Some early research in which tissue cultures were used to study disease resistance led to a view that they might provide erroneous or misleading information. For example, although major genes for resistance of potato to *Phytophthora infestans* were apparently expressed normally in cultured cells, the biochemical basis of that expression was later shown to be abnormal (Ingram, 1977). However, research into the physiology of the crown-gall disease (Butcher *et al.*, this volume) and recent studies of the expression of disease resistance to *Phytophthora parasitica* in tissue cultures of tobacco (Helgeson & Haberlach, this volume) clearly demonstrate that, if experimental systems are carefully defined, tissue cultures may provide valuable insights into a number of aspects of host-parasite interaction, including the following (Butcher, 1977; Ingram, 1977).
1. Resistance and recognition (Callow & Dow; Dixon; Helgeson &

Haberlach; this volume).
2. Tumour physiology (Buczacki; Butcher et al.; Davey et al.; Jones, this volume).
3. The mode of action of pathotoxins and enzymes.
4. Systemic infections.
5. Nutrient exchange, especially in mycorrhizal infections (Hepper & Mosse; Mason; this volume).
There are, of course, many other possibilities, but most have yet to be explored. None will provide definitive answers to the problems of host-parasite interaction, but all have the potential to provide new and significant information.

APPLICATIONS IN THE PRODUCTION OF NOVEL DISEASE-RESISTANT PLANTS

This subject has been reviewed recently by Brettell & Ingram (1979) and Thomas et al. (1979), and is referred to in a discussion by Day (this volume).

Tissue cultures form the basis of a number of techniques which have been developed to effect genetic changes in plants. Progress is already being made in the application of these techniques in the production of novel disease-resistant cultivars; in the future they could have an important role to play where conventional breeding methods have failed to produce desirable characters. Among the possibilities are the following.

It is now feasible to induce and select for mutations among popula-tions of cultured plant cells and then to regenerate plants from them (Brettell et al.; Strauss et al.; Grout & Weatherhead; this volume). This technique is likely to be particularly effective when haploid cells or protoplasts are used and where selection is for resistance to a pathotoxin such as the T-toxin of Drechslera maydis (Brettell et al., this volume). However, it is not yet known whether such procedures are widely applicable, and the nature of the genetic changes involved has yet to be determined.

The tissues of plant species which are propagated vegetatively are normally genetic mosaics with regard to many characteristics, including resistance to disease. Thus, some of the plants regenerated from the cultured cells or protoplasts of such species are more resistant than the parent plants. This phenomenon has already led to the production of new disease-resistant lines of sugar cane and potato, and its potential use for the improvement of other crops is now being assessed (Day, this volume).

Many attempts have been made to modify the genomes of cultured plant cells by means of exogenous nucleic acids (Kleinhofs & Behki, 1977). The evidence for integration and replication of this genetic material is, however, equivocal. The technique, therefore, offers no immediate prospects for the development of useful disease-resistant plants, but may be very important in the long term as methods are perfected for using plasmids and other agents as carriers of useful genes (Davey *et al.*, this volume).

Steady advances are now being made in producing somatic hybrids of crop plants by fusion of isolated protoplasts (Hanke; Day; this volume). In the long term it may be possible to use protoplast fusion to transfer desirable disease-resistance traits between related species which cannot be hybridized by conventional breeding methods.

Finally, the culture of excised embryos may now be used to grow interspecific and intergeneric hybrid plants in cases where incompatibility occurs after normal fertilization. The technique is already used by breeders in the production of disease-resistant hybrids of a number of crops (*Nicotiana* spp., barley/rye and Brassicas; Ross, this volume), and its use is likely to increase considerably in the near future. Incompatibility can also be overcome in some cases through the technique of *in vitro* fertilization. The exciting possibility of applying this technique to the generation of new disease-resistant hybrids has, however, yet to be explored.

REFERENCES

Barz W., Reinhard E. & Zenk M.H. (1977) *Plant Tissue Culture and its Biotechnological Application*. Springer-Verlag, Berlin.
Brettell R.I.S. & Ingram D.S. (1979) Tissue culture in the production of novel disease-resistant crop plants. *Biological Reviews* 54, 329-45.
Butcher D.N. (1977) Plant tumour cells. *Plant Tissue and Cell Culture* (Ed. by H.E. Street) 2nd edition, pp. 429-61. Blackwell Scientific Publications, Oxford.
Butcher D.N. & Ingram D.S. (1976) *Plant Tissue Culture*. Edward Arnold, London.
Davies D.R. & Hopwood D.A. (1980) *The Plant Genome*. The John Innes Charity, Norwich.
Dudits D., Farkas G.L. & Maliga P. (1976) *Cell Genetics in Higher Plants*. Akadémiai Kiadó, Budapest.
Durbin R.D. (1976) Nicotiana, *Procedures for Experimental Use*. United States Department of Agriculture, Washington.
Gamborg O.L. & Wetter L.R. (1975) *Plant Tissue Culture Methods*. National Research Council of Canada, Ottawa.
Gautheret R.J. (1959) *La Culture des Tisses Végétaux, Techniques et Réalisations*. Masson Cie, Paris.
Hollings M. (1965) Disease control through virus-free stock. *Annual Review of Phytopathology* 3, 367-96.
Ingram D.S. (1976) Growth of biotrophic parasites in tissue culture. *Encyclopedia of Plant Physiology: Physiological Plant Pathology* (Ed. by R. Heitefuss & P.H. Williams), pp. 743-59. Springer-Verlag, Berlin.
Ingram D.S. (1977) Applications in plant pathology. *Plant Tissue and Cell Culture* (Ed. by H.E. Street), pp. 463-500. Blackwell Scientific Publications, Oxford.

Kasha K.J. (1974) *Haploids in Higher Plants, Advances and Potential.*
  University Press, Guelph, Ontario.
Kleinhofs A. & Behki R.M. (1977) Prospects for plant genome modifica-
  tion by non-conventional methods. *Annual Review of Genetics* 11,
  79-101.
Kruse P.F. & Patterson M.K. (1973) *Tissue Culture, Methods and
  Applications.* Academic Press, New York.
Murashige T. (1974) Plant propagation through tissue culture. *Annual
  Review of Plant Physiology* 25, 135-66.
Peberdy J.F., Rose A.H., Rogers H.J. & Cocking E.C. (1976) *Microbial
  and Plant Protoplasts.* Academic Press, London.
Pierik R.L.M. (1979) In Vitro *Culture of Higher Plants.* R.L.M. Pierik,
  Wageningen.
Quak F. (1977) Meristem culture and virus-free plants. *Plant Cell,
  Tissue and Organ Culture* (Ed. by R.J. Reinert & Y.P.S. Bajaj), pp.
  598-615. Springer-Verlag, Berlin.
Reinert J. & Bajaj Y.P.S. (1977) *Plant Cell, Tissue and Organ Culture.*
  Springer-Verlag, Berlin.
Street H.E. (1974) *Tissue Culture and Plant Science.* Academic Press,
  London.
Street H.E. (1977) *Plant Tissue and Cell Culture*, 2nd edition.
  Blackwell Scientific Publications, Oxford.
Takebe I. (1975) The use of protoplasts in plant virology. *Annual
  Review of Phytopathology* 13, 105-25.
Takebe I. (1977) Protoplasts in the study of plant virus replication.
  Comprehensive Virology, 11 (Ed. by H.H. Fraenkel-Conrat &
  R.R. Wagner), pp. 237-83. Plenum, New York.
Thomas E. & Davey M.R. (1975) *From Single Cells to Plants.* Wykeham
  Publications, London.
Thomas E., King P.J. & Potrykus I. (1979) Improvement of crop plants
  via single cells *in vitro* - an assessment. *Zeitschrift für Pflanzen-
  züchtung* 82, 130.
Thorpe T.A. (1978) *Frontiers of Plant Tissue Culture, 1978.* Inter-
  national Association for Plant Tissue Culture, Calgary.
Walkey D.G.A. (1978) *In vitro* methods for virus elimination. *Frontiers
  of Plant Tissue Culture, 1978* (Ed. by T.A. Thorpe), pp. 245-54.
  International Association for Plant Tissue Culture, Calgary.
White P.R. (1963) *The Cultivation of Animal and Plant Cells.* Ronald
  Press, New York.
Willmer E.N. (1966) *Cells and Tissue in Culture, Vol. 3.* Academic
  Press, London.
Wood K.R., Boulton M.I. & Maule A.J. (1979) Application of protoplasts
  in plant virus research. *Plant Cell Cultures, Results and Perspec-
  tives* (Ed. by O. Ciferri & B. Parisi), pp. 405-10. Elsevier,
  Amsterdam.
Zaitlin M. & Beachy R. (1974) The use of protoplasts and separated
  cells in plant virus research. *Advances in Plant Virus Research* 19,
  1-35.
Zuckerman B.M. (1971) Gnotobiology. *Plant Parasitic Nematodes, Vol. 2*
  (Ed. by B.M. Zuckerman, W.F. Mai & R.A. Rhode), pp. 159-84.
  Academic Press, New York.

Section 2
Basic tissue culture methods

The culture of isolated roots

D.N. BUTCHER

Rothamsted Experimental Station, Harpenden AL5 2JQ

INTRODUCTION

Isolated roots in culture retain the organization of the intact organs
and are therefore potentially useful for the investigation of the
fundamental properties of roots. The basic methods for root culture
were developed by White (1943, 1963) and elaborated by Street and
co-workers (1957, 1966). Clones of cultured roots of the following
species have been successfully established in continuous culture:
Lycopersicon esculentum Mill., *Lycopersicon pimpinellifolium* Mill.,
Datura stramonium L., *Solanum tuberosum* L., *Helianthus annuus* L.,
Senecio vulgaris L., *Medicago sativa* L., *Trifolium repens* L., *Secale
cereale* L., *Triticum vulgare* L. cv. Hilgendorf 61, *Androcymbium
gramineum* (Cav) McBride, *Pinus serotina* Michx., *Pinus ponderosa* Dougl.
(see Butcher & Street, 1964 for references), *Brassica campestris* L. (El
Tigani, 1972) and *Isatis tinctoria* L. (Elliott & Stowe, 1971). Roots of
other species have been maintained in culture for prolonged periods but
poor lateral development has prevented multiplication of clonal material
from individual roots. Such species include *Brassica nigra* L. Koch.,
Sinapis alba L., *Raphanus sativus* L., *Callistephus hortensis* Cass.,
Acer melanoxylon R. Br., *Melilotus alba* Destr., *Pisum sativum* L., *Linum
usitatissimum* L., *Fagopyrum esculentum* Mönch., *Petunia violacea* Lindl.
(see Butcher & Street, 1964 for references) and *Comptonia peregrina* L.
(Coult.) (Goforth & Torrey, 1977). The work on the culture of isolated
roots has been reviewed by Street (1957) and Butcher & Street (1964).

   Root cultures have several characteristics which make them attractive
as host tissues for investigations of host-parasite interactions.
Firstly, roots which have a high growth rate and profuse lateral
development can provide an unlimited supply of uniform clonal material
for experiments. Secondly, isolated roots are grown under controlled
environmental conditions in the absence of contaminating micro-organisms.
Thirdly, root cultures which grow on simple nutrient media and retain
the morphology of the intact root would appear to be ideal for studying
the infection mechanisms of root pathogens.

   However, isolated root cultures have not been widely used in

Ingram D.S. & Helgeson J.P. (1980) *Tissue Culture Methods for Plant Pathologists*

investigations of root pathogens. This is mainly because roots of the
appropriate host species are often difficult to culture and many root
pathogens are facultative parasites which overgrow the culture media.
The latter does not apply to root infecting biotrophs, which require
nutrients from living tissue, and root cultures have been successfully
used to study vesicular-arbuscular mycorrhizas (Hepper & Mosse, this
volume), endoparasitic nematodes (Jones, this volume) and the clubroot
disease of crucifers (Ingram, 1969). In addition, dual cultures of
isolated roots and parasites can provide an effective way of maintaining
a continuous supply of uncontaminated pathogen.

METHODS AND MATERIALS

Culture media

The modified White's medium in Table 1 is excellent for growing roots
of *L. esculentum* (tomato) and has been widely used as the starting point
for the development of media for other species. In this version of
White's medium the iron is supplied as ferric ethylene-diaminetetra
acetate (Fe EDTA). This makes the iron available over a wide pH range
up to pH 7.5. For convenience the medium is usually prepared from the
following stock solutions.
  1. Inorganic salts (minus the ferric chloride) at 10 times the final
     strength.
  2. Iron source ($FeCl_3$ and Na EDTA) at 10 times the final strength.
  3. Vitamin stock containing thiamine hydrochloride, pyridoxine
     hydrochloride, nicotinic acid and glycine at 100 times the final
     strength. This is stored in aliquots (10 ml at -15°C).

    To prepare 1 l of medium dissolve the sucrose in 750 ml of distilled
water, add 100 ml each of stock solutions 1 and 2 and 10 ml of stock
solution 3. Adjust the pH to 4.8 using 0.1 M NaOH or 0.1 M HCl, make
up to volume and distribute to culture vessels (usually 50 ml in 100 ml
borosilicate Erlenmeyer flasks). Close flasks with cotton wool plugs
and sterilize by autoclaving at 121°C for 10-15 minutes.

    Although the medium in Table 1 supports a high growth rate of
isolated tomato roots, additional growth factors are often necessary for
other species. Roberts & Street (1955) grouped isolated roots on the
basis of their response to supplied auxin. The growth of some species
such as tomato is either not affected by or is inhibited by auxin; the
growth of others such as *S. vulgaris*, *P. sativum*, *T. vulgare*,
A. gramineum and *P. ponderosa* is increased by, and *S. cereale* is totally
dependent on, the addition of auxin. The response of isolated roots to
other growth hormones is not well documented. There is little evidence
that cytokinins stimulate root growth and in some instances they may be
inhibitory (Dropkin *et al.*, 1969). Gibberellins have only been shown to
increase the growth of tomato roots in special conditions (Butcher &
Street, 1960), and roots of certain genetic dwarfs (Mertz, 1966).

    Sucrose is the best carbon source for roots of most dicotyledonous

Table 1. Root culture medium (after White, 1943, as modified by Street & McGregor, 1952, and Sheat *et al.*, 1959).

Constituents	Mg/l
$MgSO_4.7H_2O$	731
$Na_2SO_4.10H_2O$	453
$Ca(NO_3)_2.4H_2O$	288
KNO_3	80
KCl	65
$NaH_2PO_4.2H_2O$	21.5
$MnCl_2.4H_2O$	6.0
$ZnSO_4.7H_2O$	2.65
H_3BO_3	1.50
KI	0.75
$CuSO_4.5H_2O$	0.13
H_2MoO_4	0.0017
$FeCl_3$	3.1
Sodium ethylenediaminetetra-acetate (EDTA)	8.0
Thiamine hydrochloride	0.1
Pyridoxine hydrochloride	0.1
Nicotinic acid	0.5
Glycine	3.0
Sucrose	20 000

species and is not replaceable by glucose, fructose or maltose. While 2% is the most commonly used concentration, the roots of a few species such as *P. sativum*, *Convolvulus arvensis* and *L. usitatissimum* have a higher requirement (4-6%) for optimum growth rates. In contrast, for many monocotyledonous species glucose is as good as or slightly better than sucrose as a carbon source. Thiamine is essential for the growth of many roots, while pyridoxine and nicotinic acid enhance growth but are not essential, and glycine is probably unnecessary. Recently it has been shown that the growth of some roots is significantly increased by the addition of meso-inositol (50 mg/l); for example *L. esculentum* (Sinha, 1968) and *C. peregrina* (Goforth & Torrey, 1977).

Sterilization and germination of seed

Isolated root cultures are usually initiated from root tips of aseptically grown seedlings (Fig. 1). Depending on the sensitivity of the seeds and the nature of the contaminating micro-organisms, the dry seeds are sterilized by one of the following treatments.
1. Immerse in a 1% (w/v) bromine solution for 5 min.
2. Immerse in a sodium hypochlorite solution (1% available chlorine) for 30 min.
3. Immerse in aqueous detergent (0.5% Tween 20) for 5 min and 0.1%

aqueous mercuric chloride for up to 20 min.

After sterilization wash the seeds thoroughly with sterile distilled water and place in sterile Petri dishes containing moistened filter papers. Germinate in the dark at 25-27°C and leave until the radicles or seminal roots are 20-40 mm in length.

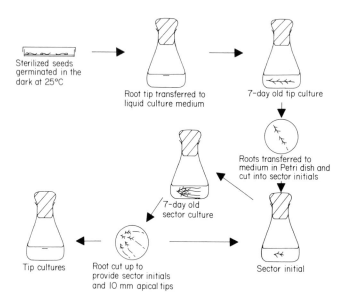

Sterilized seeds germinated in the dark at 25°C

Root tip transferred to liquid culture medium

7-day old tip culture

Roots transferred to medium in Petri dish and cut into sector initials

7-day old sector culture

Sector initial

Tip cultures

Root cut up to provide sector initials and 10 mm apical tips

Figure 1.   Procedure for culturing a clone of isolated tomato roots.

Initiation of cultures and establishment of clones

Excise apical tips (10 mm) from the sterile seedlings, transfer singly to the surface of liquid culture medium and incubate in the dark at low light intensities (less than 1000 lx) at 25-27°C. Allow the roots to grow until the main axis is 100-130 mm in length and numerous lateral roots have developed. With tomato this takes 7-10 days from the initial tips. It should be noted that roots which float on the surface of the medium usually have a higher growth rate than those which sink.

To establish a clone, excise segments of the main root axis bearing four or five newly formed lateral roots (sectors) from a single root culture and transfer individually to separate flasks of culture medium (Fig. 1). Incubate the sector cultures at 27°C until the primary lateral roots have elongated and formed secondary laterals. Use these cultures to produce new sector and tip cultures. The sectors may be used to propagate the clone and the tip cultures may be used to initiate experiments (Fig. 1).

Because the growth characteristics of roots from different species vary considerably it is often necessary to modify the duration of the incubation period to maintain a clone with a high and uniform growth rate. For less vigorous roots clonal cultures can sometimes be achieved by alternating sector and tip cultures (e.g. *I. tinctoria*, Elliott & Stowe, 1971).

REFERENCES

Butcher D.N. & Street H.E. (1960) The effects of gibberellins on the growth of excised tomato roots. *Journal of Experimental Botany* 11, 206-16.

Butcher D.N. & Street H.E. (1964) Excised root culture. *Botanical Review* 30, 513-86.

Dropkin V.H., Helgeson J.P. & Upper C.D. (1969) The hypersensitivity reaction of tomatoes resistant to *Meloidogyne incognita*: reversal by cytokinins. *Journal of Nematology* 1, 55-61.

El Tigani S. (1972) The role of plant hormones in the clubroot disease of *Brassica rapa* L. *Ph.D. Thesis, University of Cambridge*.

Elliott M.C. & Stowe B.B. (1971) Distribution and variation of indole glucosinolates in woad (*Isatis tinctoria* L.). *Plant Physiology* 48, 498-503.

Goforth P.L. & Torrey J.G. (1977) The development of isolated roots of *Comptonia peregrina* (Myricaceae) in culture. *American Journal of Botany* 64, 476-82.

Ingram D.S. (1969) Growth of *Plasmodiophora brassicae* in host callus. *Journal of General Microbiology* 55, 9-18.

Mertz D. (1966) Hormonal control of root growth. *Plant Cell Physiology* 7, 125-35.

Roberts E.H. & Street H.E. (1955) The continuous culture of excised rye roots. *Physiologia Plantarum* 8, 238-62.

Sheat D.E.G., Fletcher B.H. & Street H.E. (1959) Studies on the growth of excised roots. VIII. The growth of excised tomato roots supplied with various organic sources of nitrogen. *New Phytologist* 58, 124-41.

Sinha S. (1968) Studies on the control of secondary thickening in excised roots of *Lycopersicum esculentum* Mill. *Ph.D. Thesis, University of Glasgow*.

Street H.E. (1957) Excised root culture. *Biological Reviews* 32, 117-55.

Street H.E. & Henshaw G.G. (1966) Introduction and methods employed in plant tissue culture. *Cells and Tissues in Culture*, Vol. 3 (Ed. by E.N. Willmer), pp. 459-532. Academic Press, London.

Street H.E. & McGregor S.M. (1952) The carbohydrate nutrition of tomato roots.   III. The effects of external sucrose concentration on the growth and anatomy of excised roots. *Annals of Botany* 16, 185-205.

White P.R. (1943) *A Handbook of Plant Tissue Culture*. J. Cattell, Lancaster, Pa.

White P.R. (1963) *The Cultivation of Animal and Plant Cells*, 2nd edition. Ronald Press, New York.

Plant tissue and cell suspension culture

J.P. HELGESON

USDA, SEA, AR Plant Disease Resistance Research Unit,
Department of Plant Pathology, University of Wisconsin,
Madison, WI 53706, USA

INTRODUCTION

Throughout this book there are many examples of how cell cultures, protoplasts, excised stem tips and other tissues cultured *in vitro* can be very useful to plant pathologists. This chapter introduces the basic techniques of plant tissue and cell culture. The following are considered: media, sources and isolation of explants, physical arrangements for growing tissues, the maintenance of agar-grown and cell suspension cultures and the measurement of tissue culture growth. Many of the details given are based on experience with *Nicotiana* (tobacco) tissue cultures, but most will have general applicability. Further information may be found in the volume edited by Street (1977) and in the laboratory manual of Gamborg & Wetter (1975).

METHODS AND MATERIALS

Media

A number of good general purpose media are now available. Several of these are discussed and compared by Gamborg *et al.* (1976). The one developed by Murashige & Skoog (1962) and altered slightly by Linsmaier & Skoog (1965) is useful for many purposes (Table 1). The inorganic constituents are satisfactory for most purposes, but modifications of the organic constituents are commonly needed.

The stock solutions for the Linsmaier & Skoog (LS) medium are given in Table 1. To prepare the kinetin solution add about 10.15 mg of kinetin to 800 ml of distilled water. Heat gently with stirring to dissolve, and filter the solution through Whatman No. 1 filter paper. Read the absorbance of the solution at 267 nm in a spectrophotometer and adjust the volume so that the absorbance is 0.94. For 50 μm solutions of other cytokinins use the following values at pH 7 for aqueous solutions: benzyladenine, 0.84 at 268 nm; zeatin, 0.808 at 267 nm; dimethylallylaminopurine, 0.99 at 268 nm.

Ingram D.S. & Helgeson J.P. (1980) *Tissue Culture Methods for Plant Pathologists*

Table 1. Constituents of Linsmaier & Skoog (LS) medium

INORGANIC SALTS STOCK SOLUTIONS

Bottle	Constituent	Stock solution (g/l)	Use (ml/l)	Final concentration in medium (mg/l)
A	NH_4NO_3	82.5	20	1650
B	KNO_3	95.0	20	1900
C	$CaCl_2.2H_2O$	88.0	5	440
D	KH_2PO_4	34.0	5	170
E	H_3BO_3	1.24	5	6.2
	KI	0.166		0.83
	$Na_2MoO_4.2H_2O$	0.05		0.25
	$CoCl_2.6H_2O$	0.005		0.025
F	$MgSO_4.7H_2O$	74.0	5	370.0
	$MnSO_4.4H_2O$	4.46		22.3
	$ZnSO_4.7H_2O$	1.72		8.6
	$CuSO_4.5H_2O$	0.005		0.025
G	$FeSO_4.7H_2O$	1.393	20	27.86
	Na_2EDTA	1.863		37.26

ORGANIC MATERIALS (added to the medium at the concentration indicated)

Sucrose	30 g/l
Myo-inositol	100 mg/l
Kinetin	21.5–2150.0 μg/l*
Thiamine HCl	0.4 mg/l
Indole 3-ylacetic acid	0.175–2.0 mg/l*
Agar	6–10 g/l

---

*Concentrations of these phytohormones, or others used as substitutes, are varied according to the tissue used and the morphology of tissue desired.

To prepare 1 l of LS medium with 11.5 μM indole 3-ylacetic acid (IAA) and 1 μM kinetin, proceed as follows: to 500 ml of distilled water in a 1 l Erlenmeyer flask containing a magnetic stirring bar add 30 g of sucrose, 100 mg of myo-inositol and 0.4 mg of thiamine. Then add 20 ml each of stock solutions A, B and G and 5 ml each of stock solutions C, D, E and F (Table 1). Also add 20 ml of the kinetin stock solution.

In a small beaker dissolve 2 mg of IAA in *c.* 1 ml of ethanol, add the solution quickly to the stirring medium and rinse the beaker with distilled water. Bring the volume of the medium to about 950 ml with distilled water and adjust the pH to 5.8 ± 0.2. Finally, add the contents of the flask to a graduated cylinder and adjust the volume to 1 l.

If agar-gelled medium is desired add 10 g of agar (preferably Difco Bacto-agar No. 4001). When pre-sterilized plastic containers are used autoclave the medium in a Pyrex flask (twice the volume of the medium) for 15 min at 121°C and then dispense the sterile medium into the pre-sterilized containers. (At this step, care should be taken to prevent spills caused by superheating; do not shake the medium immediately upon removal from the autoclave.) When glass tissue culture vessels are used melt the agar in the medium by heating for 20-30 min at 100°C, stir the solution vigorously, dispense the final desired volumes into the containers and sterilize medium and containers for 15 min at 121°C.

For cell suspension cultures add the medium without agar to the culture flasks; to provide proper aeration it is generally best to add a volume of medium no greater than one fifth of the flask. Plug the flask with cotton wool and autoclave for 15 min at 121°C.

At times the addition of heat-labile materials may be required. These can be sterilized by passing a solution through a membrane filter of 0.22 μm pore size. Alternatively the DMSO procedure described by Schmitz & Skoog (1970) may be used. Such materials can be added to the tissue culture medium after the medium has cooled, but before it has gelled. (A water bath at 45° can prevent gelation of Difco Bacto-agar No. 4001.) With cell suspension media the additions can be made at any time prior to inoculation of the medium with cells. Use freshly prepared media whenever possible. However, if medium is to be stored, protect it from light.

Sources and isolation of tissue explants

Tissue cultures can be derived from many parts of the plant. For example pith tissues, root tips and anthers can all be cultured *in vitro*. Obviously the source of the explant will be dictated by the particular research. Thus, to rid plants of virus the extreme limits of the plant apex are cultured and differentiated into new plants; to test for cytokinin activity tissues such as tobacco pith are used (Skoog *et al.*, 1967). The isolation and culture of tobacco pith tissues is described by Helgeson & Haberlach (this volume).

Regardless of the source of the tissue, certain special precautions must be taken to ensure that the surface of the explant is free from contaminating micro-organisms. One useful surface sterilization procedure is to dip the explants into 70% ethanol for 30 s and then soak them in 1% (w/v) sodium hypochlorite; such a solution can be made by diluting ordinary household bleach with water (1:5). After the hypochlorite treatment rinse the explants 2-3 times with sterile distilled water. Certain other tissues, and particularly seeds, may require more rigorous procedures. Flaming the surface or two separate soakings in hypochlorite may be required. Generally, to avoid damage to tissues, it is best to use the least stringent method that will successfully free an explant of micro-organisms.

Sterile roots are obtained conveniently by excising root sections from seedlings grown from seeds germinated under aseptic conditions on agar-solidified LS medium lacking phytohormones. Alternatively the shoots from many herbaceous plants may be excised, surface sterilized and then planted in sterile culture medium lacking phytohormones or containing a small amount of auxin (*c*. 0.1 mg/l of indole-3-butyric acid). The roots that arise are excised and cultured separately.

Growing the cultures

Growth on solid media   The growth of tissue on a solid medium merely requires placing the explant on the medium. It is usually best if the cut side of the explant is placed in contact with the medium. Good adhesion of the tissue to the medium is usually obtained even without pressing the explant into the medium, especially with freshly prepared media with a film of moisture on the surface.

Cell suspension cultures   Although tissues growing on agar-solidified medium may be quite firm, those grown in suspension cultures may be loose and friable. Suspension cultures may be started from a piece of tissue from an agar-grown culture specifically plated onto a medium to induce friable growth (e.g., with tobacco, the combination of 11.5 µM IAA and 0.1 µM kinetin will result in such tissue). A piece of tissue (*c*. 1-2 g) is dropped into the suspension culture flask. After a day or two the agitation of the shaker will usually disperse the cells into small clumps. The clumps may be kept relatively small by filtering every other transfer through nylon or stainless steel mesh of the appropriate size (*c*. 350-400 µ mesh is useful for tobacco or maize cultures).

Measuring cell growth   It is of critical importance to log the course of the growth of cultures. With suspension cultures and loose friable tissues, for example, stock lines are maintained best if they are transferred during the exponential phase. With tobacco, doubling times for tissue on agar medium may be 84-96 h, but the same cultivar may grow twice as fast in suspension culture.

With agar-grown cultures removing pieces at various intervals and determining their fresh or dry weight will give a reasonable idea of how such tissues are growing. With suspension cultures aliquots can be removed with a wide-bore pipette, and either weights or packed cell volumes can be determined. A Cornwall syringe-pipette with a very wide bore (8-12 needle) is convenient for transferring measured volumes and extracting measured volumes for growth measurements. A further description of growth measurements and a computer programme for calculations can be found in Helgeson (1979).

Maintaining stock lines   Callus tissue growing on explants can be excised and maintained in culture. To accomplish this excise the callus, cut it into 30-50 mg pieces (fresh weight) and transfer the pieces to newly prepared medium. Make subsequent transfers by cutting up the callus from the previous transfer and transferring the pieces to new medium. It is best to select healthy portions of the old callus for transfer and to transfer tissue that is still actively growing. The frequency of the transfer will, of course, depend on the growth rate of the tissue. For loose friable tissues transfers while tissues are still in the exponential phase will generally give best results. With tobacco the exponential phase extends up to about half-maximal growth. With some compact slow-growing tissues it may be desirable to transfer only the tissues from the leading edge of the callus and to discard the central portion.

In the case of suspension cultures the volume of inoculum is often critical, since tissues can go through a long lag phase if the amount of inoculum is low relative to the total volume of the medium. This lag can be avoided by transferring 1 part of old medium with cells to a flask containing 4-5 parts of fresh medium. Cells so transferred should be in the exponential phase, and weekly transfers may be required to keep them in this state. The Cornwall pipette is useful for transfers; alternatively, the tip of a 5 or 10 ml pipette can be broken off to facilitate transfers from thick suspensions of cells.

Conditions for growing tissues

Containers   A large number of different tissue culture containers are available. Many are made of clear plastic and are pre-sterilized; however, standard glass Petri dishes and flasks are equally suitable. Petri dishes, preferably the 20 x 100 mm size, are particularly convenient if tissue will grow well in darkness, since they can be stacked to permit large stock plantings in a relatively small space. Glass Erlenmeyer flasks plugged with bacteriological cotton wool are quite suitable for batch suspension cultures.

Growth chambers and incubators   Unless light is required by the tissues, standard bacteriological incubators are satisfactory. However, if the cultures grow better in light, the commercially available refrigerated and lighted incubators are usually satisfactory since tissues often require only low levels of light. If air flow within the chamber is too high it is useful to filter the air or to seal the tissue culture containers with Parafilm. Generally it is not desirable to grow tissues in standard growth chambers together with potted plants, since insect infestations of cultured tissues are likely to occur.

Shakers   Cell suspension cultures require constant shaking. Healthy tissue can completely exhaust the oxygen in stationary flasks within a few minutes. For proper aeration gyrorotary shakers are more satisfactory than reciprocal shakers. A rate of 120 cycles/min with a 5 cm throw will give good aeration even of cultures grown in a 2.8 l Fernbach flask.

Light and temperature regimes   Plant tissue cultures from many species grow well between 25° and 28°C. Some, however, may require temperatures of 20-24°C. In general, temperatures above 30°C or below 20°C should be avoided unless long term, slow growth for storage is desired (see Henshaw *et al.*, this volume).

It is worth noting that the heat capacity of water in the medium is very large and is sufficient to minimize quite wide oscillations in temperature within the incubator. For example, I have measured fluctuations of less than ± 0.5° inside Petri dishes when the air in the chamber was fluctuating at ± 2°. Thus, if substantial volumes of medium are used, the temperature control inside vessels stored in even relatively inexpensive incubators will be very good.

Although many tissues grow well in the dark, it appears that some tissues prefer light (e.g. potato tissues prefer 1000-2000 lx). This lighting can be continuous or on a regular photoperiod, but it should not be random.

Inducing cultured cells and tissues to differentiate roots or shoots

Cultures from some species will differentiate roots or shoots if supplied with appropriate balances of cytokinin and auxin (Skoog & Miller, 1957). For example, tobacco tissue cultures grown on LS medium containing 1 µM IAA and 10 µM kinetin, benzyladenine, zeatin or dimethylallyladenine will often produce shoots. Similarly, the medium of Murashige & Skoog (1962) supplemented with 2.2 µM zeatin and 0.5 µM IAA is effective for producing shoots from callus tissues from several potato varieties (Campbell & Helgeson, unpublished data). When differentiation of tissue is desired a 16 h photoperiod may be very beneficial (Murashige, 1974).

REFERENCES

Gamborg O.L., Murashige T., Thorpe T.A. & Vasil I.K. (1976) Plant tissue culture media. *In Vitro* 12, 473-8.
Gamborg O.L. & Wetter L.R. (1975) *Plant Tissue Culture Methods*. National Research Council of Canada, Ottawa.
Helgeson J.P. (1979) Tissue and cell suspension culture. *Nicotiana: Procedures for Experimental Use. USDA Technical Bulletin 1586* (Ed. by R.D. Durbin), pp. 52-9. US Government Printing Office, Washington.
Linsmaier E.M. & Skoog F. (1965) Organic growth factor requirements of tobacco tissue cultures. *Physiologia Plantarum* 18, 100-27.
Murashige T. (1974) Plant propagation through tissue cultures. *Annual Review of Plant Physiology* 25, 135-66.
Murashige T. & Skoog F. (1962) A revised medium for rapid growth and bioassays with tobacco tissue cultures. *Physiologia Plantarum* 15, 473-97.
Schmitz R.Y. & Skoog F. (1970) The use of dimethylsulfoxide as a solvent in tobacco bioassays of cytokinins. *Plant Physiology* 45, 537-8.

Skoog F., Hamzi H.Q., Szweykowska A.M., Leonard N.J., Carraway K.L., Fujii T., Helgeson J.P. & Leoppky R.N. (1967) Cytokinins: structure/ activity relationships. *Phytochemistry* 6, 1169-92.
Skoog F. & Miller C.O. (1957) Chemical regulation of growth and organ formation in plant tissues cultured *in vitro*. *Symposia of the Society for Experimental Biology* 11, 118-31.
Street H.E. (1977) *Plant Tissue and Cell Culture*, 2nd edition. Blackwell Scientific Publications, Oxford.

The preparation, manipulation and culture of plant protoplasts

D.E. HANKE

Botany School, Downing Street, Cambridge CB2 3EA

INTRODUCTION

The first protoplasts were produced in 1892 by the mechanical method. With this procedure strongly plasmolysed tissue is finely chopped and then slightly deplasmolysed. Intact protoplasts emerge from cells in which cuts have been restricted to the space between the cell wall and the protoplast. Yields are low, the method is restricted to tissues of long, tubular, highly vacuolate cells (e.g. onion scales) and hyperplasmolysis is known to be deleterious. However, the mechanical method is the only way to prepare protoplasts which have not been exposed to culture filtrates of wood rotting fungi (Prat & Roland, 1970).

The use of enzymes to remove the cell wall began in 1960 when Cocking released protoplasts by treating tissues with a preparation from *Myrothecium verrucaria*. The medium was buffered osmotically to prevent bursting of the protoplasts, but in addition mild plasmolysis facilitated their release. Commercially available cellulases and pectinases were first used successfully by Takebe in 1968 (Cocking, 1972).

At first protoplasts in culture would regenerate a cell wall (it is difficult to stop them) but do little else. The secret of obtaining cell divisions and regeneration was found to be a high protoplast density of $>10^6$ protoplasts per ml (Ruesink, 1971; Cocking, 1972). More recently regeneration has been achieved from low densities in more complete media which include sources of organic nitrogen (Bui-Dang-Ha & Mackenzie, 1973; Kao & Michayluk, 1975); it appears that freshly prepared protoplasts leak amino acids and other metabolites. Plasmolysis of tissues has been shown to increase the activities of proteases and nucleases, while treatment with cellulase causes the plasma membrane to become leaky (Taylor & Hall, 1976).

Early fusion methods used high Na^+ (0.25 M; Carlson et al., 1972), but rates of survival were low and this technique has been superseded by incubation at 37°C in a medium at pH 10.5 containing 50 mM $CaCl_2$ (Keller & Melchers, 1973), or by treatment with polyethylene glycol to agglutinate the protoplasts which then fuse when this substance is

Ingram D.S. & Helgeson J.P. (1980) *Tissue Culture Methods for Plant Pathologists*

washed away (Kao et al., 1974).

Protoplasts have already proved a useful tool for plant pathologists. Most spectacularly, they have revolutionized plant virus studies, enabling for the first time simultaneous inoculation of large numbers of cells which then support a synchronized infection yielding, eventually, high virus titres (Zaitlin & Beachy, 1974). This system has been used to investigate, for example, the synthesis of host- and virus-specific proteins during infection (Siegel et al., 1978), and the causes of resistance (Zaitlin & Beachy, 1974) and necrosis (Koike et al., 1977).

Protoplasts are undoubtedly the best experimental system for investigating effects at the outer surface of the plasma membrane, and have already proved useful in studies of the action of host-specific toxins (Samadder & Scheffer, 1968; Earle et al., 1978). In this context Earle et al. proposed exploiting fusion techniques, e.g. by fusing protoplasts and mitochondria from race N and race T of maize. Future work with protoplasts will involve closer study of species- or race-specific properties of the plasma membrane surface (Hanke, 1979) which are almost certainly involved in host-specific interactions with pathogens (e.g. the hypersensitive response and biotrophy).

METHODS AND MATERIALS

All operations should be performed aseptically.

Log phase cultures of rapidly growing callus cells in suspension and mesophyll tissue of young, rapidly growing leaves are the tissues of choice. In these tissues almost the whole surface of every cell can be subjected to simultaneous enzymatic attack. Any check to growth lowers protoplast yields, probably by increasing the thickness of cell walls (Uchimiya & Murashige, 1974; Watts et al., 1974). I grow leaves for protoplasts in still air at high humidity between 26 and 30°C, surface sterilize them by immersion for 15 min in a 2% (w/v) sodium hypochlorite solution, wash them in three changes of sterile distilled water, and cut them transversely into strips 1 mm wide immediately before use. Suspension culture cells have the advantage that they are already aseptic and that I know exactly what medium to use for culture. However, leaf mesophyll tissue is often more convenient to obtain and has so far proved easier to regenerate into plants.

Prior to adding the enzymes, plasmolyse the tissue for 10 min in an osmoticum of 0.35 M sorbitol and 4 mM $CaH_4(PO_4)_2$ in glass distilled water. This treatment increases the quality and life of the resultant protoplasts. It seems that as each protoplast shrinks on plasmolysis the surplus plasma-membrane appears as cytoplasmic vesicles full of the extracellular medium; since these vesicles persist, their formation during enzyme treatment is harmful.

Next, transfer the tissues to a solution of enzymes in the osmoticum. For tissues derived from monocotyledonous plants use 5% w/v Onozuka SS

and 1% w/v Macerozyme, with the pH adjusted to 5.0 using HCl/NaOH.
(Onozuka SS and Macerozyme are made by the Yakult Manufacturing Co.,
Tokyo, and supplied in UK by R.W. Unwin & Co. Ltd, 12 Prospect Place,
Welwyn, Herts.) For tissues derived from dicots use 5% w/v Driselase
(Fluka AG, Buchs SG; supplied in UK by Uniscience Ltd, 8 Jesus Lane,
Cambridge) and 0.5% w/v desalted pectinase (Sigma Chemical Co., London)
with the pH buffered at 5.7 using 10 mM MES/NaOH. To desalt Sigma
pectinase prepare a 20% w/v solution of the enzyme in water and stir it
overnight at 4°C. Centrifuge this solution and pass the supernatant
down Sephadex G25. Use a lyophilysate of the high molecular weight
fraction (i.e. the fraction which is totally excluded from the gel phase
of Sephadex G25). Filtering the final solution of enzymes in osmoticum
through a Millipore filter, pore size 0.45 μm, before use results in an
improvement in protoplast quality.

Use 0.3-0.4 g fresh weight of tissue in 4 ml of enzyme solution and
incubate in lidded, disposable plastic dishes at 25°C on a small orbital
shaker at 45 rev/min for the first hour and 20 rev/min thereafter. For
sampling the digest, and also for moving protoplast suspensions from one
vessel to another, use Pasteur pipettes cut off to give a wide (4 mm)
mouth to eliminate shear forces. For observation, samples should be
examined under the microscope in chambers made from thin strips of cover-
slip, cut with a diamond point and cemented flat to a microscope slide
with DPX mountant (Griffin & George Ltd, Gerrard Biological Centre, East
Preston BN16 1AS), topped with a large cover-slip.

Digestion is complete within 2-6 hours. Filter the digest slowly
through three layers of cotton muslin to sieve off cuticle and cell
clumps and to adsorb pieces of cell wall. Remove the enzymes either by
repeated (3x) centrifugation for 2 min at 45 g and resuspension in fresh
osmoticum, or by filtration under gravity in a folded Whatman GF/C glass
fibre paper circle in a glass funnel while osmoticum, 100 ml for 4 ml of
digest, is gently poured over the protoplasts (Hanke & Northcote, 1974).
Although centrifugation is quicker and easier, the condition of the
protoplasts is not as good as after filtering. The most exacting
criterion of good condition for protoplasts from suspension culture
cells is vigorous cytoplasmic streaming as seen using Nomarski optics.
A rather less exacting criterion is the profusion of cytoplasmic strands
which is absent from poor quality protoplasts. When the peripheral
chloroplasts obscure what goes on inside mesophyll protoplasts, low
frequency of bursting indicates good quality.

Callus protoplasts should be cultured in their original growth medium
supplemented with 0.25 M sorbitol, and mesophyll protoplasts in the
medium of Uchimiya & Murashige (1976). Protoplasts are easily burst by
physical disturbances and should always be moved gently and slowly.
They are very sensitive to detergents and care must be taken that
culture vessels have been thoroughly rinsed. The surface properties of
protoplasts freshly isolated from rapidly growing callus cells change
within 30-60 min of washing away cellulase, consistent with very rapid
synthesis of surface glucan (Hanke, 1979) and fusion is only possible
before this transition. A 10 min treatment with dilute (0.1%) Onozuka

SS solution restores the fusibility of freshly isolated protoplasts.
High pH (Keller & Melchers, 1973) or a combination of high pH and
polyethylene glycol (Grimsley et al., 1977) are the most satisfactory
methods for achieving protoplast fusion and recovering live fusion
products. Using polyethylene glycol, fusion is extensive but viability
of fusion products is poor. Recovery from high pH treatment is excellent, but fusions are less frequent. When protoplasts are treated to
induce fusion after they have settled to the bottom of the dish large
scale multicellular fusions are the rule and the giant multinucleate
products are fragile and do not survive in culture. Treatment while
protoplasts are kept in suspension results in a larger proportion of
aggregates with only two protoplasts. To obtain fusion suspend the
protoplasts at 10^7 per ml with gentle shaking for 40 min in a 50% w/v
polyethylene glycol (molecular weight 6000), 50 mM $CaCl_2$ solution.
During the next 40 min, while continuing to shake gently, dilute the
polyethylene glycol gradually with a 0.30 M sorbitol, 50 mM $CaCl_2$, 50
mM glycine/NaOH solution (pH 10.5) up to 30 times the volume of the
original polyethylene glycol-containing solution. Collect the fusion
products by centrifugation or filtration and resuspend in the medium
for culture (Grimsley et al., 1977).

Protoplasts of callus origin should be transferred to sorbitol-free
growth media after 2 days, and protoplasts of mesophyll origin after 5
days, since unnecessarily prolonging the period in osmotically buffered
solutions inhibits regeneration. Before transfer a small sample of
protoplasts should be tested for osmotic stability by suspending in the
new medium. Do not dilute protoplast suspensions below about 10^6 per ml
of new medium, since culture at low density often leads to the production of misshapen (budded) multinucleate cells. This abnormality is
due to the failure to reform a complete wall when all the matrix
component has been released to the medium (Hanke & Northcote, 1974).
Budded protoplasts never regenerate.

REFERENCES

Bui-Dang-Ha D. & Mackenzie I.A. (1973) The division of protoplasts from
    Asparagus officinalis L. and their growth and differentiation.
    Protoplasma 78, 215-21.
Carlson P.S., Smith H.H. & Dearing R.D. (1972) Parasexual interspecific
    plant hybridization. Proceedings of the National Academy of Sciences,
    USA 69, 2292-4.
Cocking E.C. (1972) Plant cell protoplasts - isolation and development.
    Annual Review of Plant Physiology 23, 29-50.
Earle E.D., Gracen V.E., Yoder O.C. & Gemmill K.P. (1978) Cytoplasm-
    specific effects of Helminthosporium maydis Race T toxin on survival
    of corn mesophyll protoplasts. Plant Physiology 61, 420-4.
Grimsley N.H., Ashton N.W. & Cove D.J. (1977) Complementation analysis
    of auxotrophic mutants of the moss, Physcomitrella patens, using
    protoplast fusion. Molecular and General Genetics 155, 103-7.
Hanke D.E. (1979) Plasma-membrane surface components investigated
    using protoplasts. Plant Organelles. Methodological Surveys (B)

Biochemistry, Volume 9 (Ed. by E. Reid), pp. 196-8. Ellis
  Horwood Ltd, Chichester.
Hanke D.E. & Northcote D.H. (1974) Cell wall formation by soybean
  callus protoplasts. *Journal of Cell Science* 14, 29-50.
Kao K.N., Constabel F., Michayluk M.R. & Gamborg O.L. (1974) Plant
  protoplast fusion and growth of intergeneric hybrid cells. *Planta*
  120, 215-27.
Kao K.N. & Michayluk M.R. (1975) Nutritional requirements for growth
  of *Vicia hajastana* cells and protoplasts at a very low population
  density in liquid media. *Planta* 126, 105-10.
Keller W.A. & Melchers G. (1973) The effect of high pH and calcium on
  tobacco leaf protoplast fusion. *Zeitschrift für Naturforschung* 28c,
  737-41.
Koike M., Hibi T. & Yora K. (1977) Infection of cowpea mesophyll
  protoplasts with cucumber mosaic virus. *Virology* 83, 413-6.
Prat R. & Roland J.-C. (1970) Isolement mécanique et étude ultrastruc-
  turale préliminaire de protoplastes végétaux. *Comptes rendus
  hebdomadaire des séances de l'Académie des sciences* 271D, 1862-5.
Ruesink A.W. (1971) Protoplasts of plant cells. *Methods in Enzymology
  Volume XXIII* (Ed. by S.P. Colowick & N.O. Kaplan), pp. 197-209.
  Academic Press, London & New York.
Samadder K.R. & Scheffer R.P. (1968) Effect of the specific toxin in
  Helminthosporium victoriae on host cell membranes. *Plant Physiology*
  43, 21-8.
Siegel A., Hari V. & Kolacz K. (1978) The effect of tobacco mosaic
  virus infection on host and virus-specific protein synthesis in
  protoplasts. *Virology* 85, 494-503.
Taylor A.R.D. & Hall J.L. (1976) Some physiological properties of
  protoplasts isolated from maize and tobacco tissues. *Journal of
  Experimental Botany* 27, 383-91.
Uchimiya H. & Murashige T. (1974) Evaluation of parameters in the
  isolation of viable protoplasts from cultured tobacco cells. *Plant
  Physiology* 54, 936-44.
Uchimiya H. & Murashige T. (1976) Influence of the nutrient medium on
  the recovery of dividing cells from tobacco protoplasts. *Plant
  Physiology* 57, 424-9.
Watts J.W., Motoyoshi F. & King J.M. (1974) Problems associated with
  the production of stable protoplasts of cells of tobacco mesophyll.
  Annals of Botany 38, 667-71.
Zaitlin M. & Beachy R.N. (1974) Protoplasts and separated cells: some
  new vistas for plant virology. *Tissue Culture and Plant Science*
  (Ed. by H.E. Street), pp. 265-85. Academic Press, London & New York.

Guidelines in the culture of pollen *in vitro*

N. SUNDERLAND

John Innes Institute, Colney Lane,
Norwich NR4 7UH

INTRODUCTION

Three procedures are available for diverting pollen from its normal role as the male gametophyte: anther culture, inflorescence culture and pollen culture.

Anther culture is the oldest and most widely practised; introduced by Guha & Maheshwari (1964) on the Solanaceous species *Datura innoxia*, the technique has been applied with varying degrees of success to many plants including important crops such as rice and wheat (Hu, 1978), barley (Mix *et al.*, 1978), rye (Wenzel *et al.*, 1977), maize (Miao *et al.*, 1978), brassicas (Keller & Stringham, 1978), potato (Sopory *et al.*, 1978) and tobacco (Sunderland & Dunwell, 1977). The aim is to find a protocol which permits development of the pollen with minimal or no growth of the surrounding somatic tissues. The pollen develops inside the anthers as organized embryonic structures (tobacco, brassicas) or as masses of unorganized callus cells (grasses).

Inflorescence culture is a recent, labour-saving modification, described by Wilson (1977) for barley. Again the culture procedure is such as to favour growth of the pollen. The technique is more limited in scope than anther culture but deserves consideration for use on other grasses and species that produce small flowers and reduced perianths.

Pollen culture, dating from 1974 (Nitsch), has been realized only in a few Solanaceous species. A period of anther preculture is necessary to ensure growth of the pollen after isolation, otherwise it degenerates rapidly. Isolation can be circumvented by floating anthers on the surface of shallow layers of liquid medium. The anthers dehisce while floating and pollen shed into the medium develops free of the anther tissue (Sunderland & Roberts, 1979).

The pollen is responsive to culture only during a short period in the early development of the anthers. This begins just after the release of the microspores from the meiotic tetrads (note: in some species the tetrad configuration is retained) and ends just after completion of the

first mitotic division which initiates the formation of the male
gametophyte (Sunderland & Dunwell, 1977). Within this period, there is
an optimal stage for each species or cultivar. Some respond best before
the first mitotic division (class I: pre-mitotic, e.g. grasses), some
when the division has begun (class II: mitotic, e.g. tobacco) and others
just after completion of the division and formation of the generative
cell (class III: post-mitotic, e.g. *Nicotiana sylvestris*).

Pollen plants show a wide range of ploidies. Non-haploids arising
spontaneously from haploid cells carry genes all in the homozygous state.
Homozygous diploids can also be generated from haploids by colchicine
and other treatments (Jensen, 1974). These doubled haploids are
valuable in plant improvement and combination breeding (Nitzsche &
Wenzel, 1977). (Note: non-haploids can arise from unreduced pollens
formed by irregular cleavage at meiosis; these non-haploids may display
heterozygosity. There are genotypic differences in the occurrence and
frequency of irregular pollens; the frequency increases with plant age
(Sunderland & Dunwell, 1977)). Haploids also have potential in mutation
breeding and genetic research. They have been little exploited in plant
pathology, but offer scope in the isolation of pathogen-resistant
strains (Carlson, 1973; Strauss *et al.*, this volume), and in the
eradication of viruses (Maia *et al.*, 1975) and other pathogens which are
not pollen transmitted (this applies to the crown-gall organism).

GENERAL PROCEDURE

Harvesting of flower buds or inflorescences

Individual flower buds, parts of inflorescences, whole inflorescences or
tillers are harvested depending on the species, the morphological
complexity of the inflorescence and its position in the plant.
Selection is aided by reference to criteria such as bud length, corolla
length and the length of the emerging flag leaf or spike. Wherever
possible, material should be harvested near the beginning of the
flowering period. The response of the pollen to culture decreases with
plant age. This response is also influenced by photoperiod, temperature
and the nutrient status of the plants (Sunderland & Dunwell, 1977;
Sunderland, 1978).

Surface sterilization

The flower buds are closed and sometimes surrounded by parental tissue
(barley) at the critical stage. Standard hypochlorite solutions (1%
available chlorine) containing a wetting agent usually suffice to remove
surface organisms. In barley, an alcohol swab of the ensheathing flag
leaf is sufficient. Where buds are well protected by scales (e.g. peony)
dipping and flaming in alcohol is best.

Selection of anthers at the optimal developmental stage

For maximal culture efficiency inocula should be as close as possible

to the optimal stage. In morphologically identical buds, spikes or
tillers, the stage can differ sufficiently to give an all-or-none
response. Cytological examination of the pollen is the surest way of
achieving uniformity. One or more test anthers are removed aseptically
from the harvested material and crushed gently in acetocarmine; the
stained pollen is examined in the light microscope.

In class I plants the test anther should show microspores having a
well developed exine, a vacuole and a small nucleus lying in a thin
layer of peripheral cytoplasm. Class II anthers show varying proportions of spores in division whereas in class III the pollen should
consist entirely of young bicellular gametophytes showing a small
generative cell either attached to or detached from the inner pollen
wall. Dense staining of the cytoplasm may obscure the other nucleus.
Starch deposition should not have begun. (Note: *D. innoxia* should
be regarded as class I if only haploids are required - otherwise this
species is class II. *Hyoscyamus* spp. are class I without pretreatment
but class II with pretreatment (Sunderland & Wildon, 1979)).

The degree of uniformity that can be attained both between and within
cultures varies from one species to another. In species having small
and determinate numbers of anthers asynchrony within a bud is relatively slight. Asynchrony increases with increasing numbers of anthers per
bud. Similarly in grasses, synchrony between spikelets in the same
spike varies with the species. In maize and sorghum, where inflorescences are large and morphologically complex, there may be wide asynchrony.
In barley, on the other hand, asynchrony is relatively slight and
anthers from several spikelets in the centre of the spike can be pooled
without unduly affecting uniformity. In compound grass spikelets only
one or two florets will be at the desired stage, but again anthers from
several spikelets can be safely pooled. The position of the desired
spikelets or florets in any spike changes with plant age.

Pretreatment of the staged material

Pretreatment is a means of inducing pollen into a state of incipient
embryogenesis before culture. Temperature stress is the most effective
procedure so far devised. The staged floral parts are kept in sealed
containers in darkness. Water included to maintain humidity should not
be in contact with the tissues. In barley removal of the awns and the
upper- and lower-most spikelets is advisable. The de-awned spikes are
kept in sterile Petri dishes sealed with Parafilm. In grasses in which
the spike has already emerged at the critical stage the excised spikes
(after removal of the test spikelet) can be wrapped in polyethylene,
then in thin aluminium foil, to minimize water loss. In this case
sterilization is delayed until after the pretreatment.

Cold temperatures (3-6°C) have been mostly used (Nitsch, 1974), but
any temperature up to about 25°C will suffice. There is evidence that
warm temperatures, in the range 10-16°C, may be more effective for some
species (Sunderland & Wildon, 1979). The time of pretreatment is
critical in relation to the initial anther stage. The most effective

combination of time, temperature and stage has to be determined for every species or cultivar. At 4-5°C about 21 days may be required for the optimal effect in some grasses, but this treatment is not universally effective. The minimal universal temperature is 7°C, at which about 14 days' pretreatment is required. At 15°C periods of about 7 days should be tried, and at 25°C periods of 3 days; the time requirement becomes progressively more critical with increasing temperature.

Preparation of cultures

For inflorescence culture (barley) the de-awned spikes are removed from the pretreatment and shaken individually in liquid medium (10-15 ml) in plugged Erlenmeyer flasks (250 ml).

For anther culture anthers are dissected aseptically and plated on agar or agar/charcoal media. In tobacco there is an optimal relationship between the number of anthers used, the volume of medium and the size of the culture vessel (Dunwell, 1979). Alternatively, excised anthers should be floated on liquid medium (4-5 ml) in deep plastic Petri dishes (50 x 20 mm). The dishes are sealed with Parafilm and kept stationary.

Table 1. Potato media for cereal anther culture (quantities in mg/l except where stated), from Chuang *et al.*, 1978.

	PM1 culture	PM2 regeneration	PM3 maintenance
KNO_3	1000	500	100
NH_4NO_3	–	850	–
$(NH_4)_2SO_4.H_2O$	100	–	–
$Ca(NO_3)_2.4H_2O$	100	–	200
$CaCl_2.2H_2O$	–	200	–
KCl	35	–	–
$MgSO_4.7H_2O$	125	200	450
KH_2PO_4	200	100	–
FeEDTA	$10^{-4}M$	$10^{-4}M$	$1-3 \times 10^{-4}M$
Thiamin HCl	1.0	1.0	–
Sucrose	90 000	30 000	15 000
Kinetin	0.5	–	–
2,4-D	2.0	–	–
Potato extract	10%	5%	5%

pH is adjusted to 5.8 before medium is autoclaved.
Potato extract: diced potato tuber 100 000 boiled in distilled water, strained and cooled. For 5% use half strength.
Agar used at 7000 to 8000.

Table 2. Synthetic media for use in anther culture (quantities in mg/l).

	Cereals (Chu, 1978)	Solanaceae (Sunderland & Roberts, 1979)		Brassicas (Keller & Stringham, 1978)
KNO_3	2830	950		2500
NH_4NO_3	-	825		-
$(NH_4)_2.SO_4$	463	-		134
$CaCl_2.2H_2O$	166	220		750
$MgSO_4.7H_2O$	185	185		250
KH_2PO_4	400	85		-
$NaH_2PO_4.H_2O$	-	-		150
Sequestrene Fe 330	-	-		40
$FeSO_4.7H_2O$ ⎫ mixed in water before	27.8	27.8		-
Na_2EDTA ⎭ adding to medium	37.3	37.3		-
$MnSO_4.4H_2O$	4.4	11.2 ⎫		10
H_3BO_3	1.6	3.1		3.0
$ZnSO_4.7H_2O$	1.5	4.3		2.0
KI	0.8	0.4		0.8
$Na_2MoO_4.2H_2O$	-	0.13		0.25
$CuSO_4.5H_2O$	-	0.013 ⎬optional		0.025
$CoCl_2.6H_2O$	-	0.013		0.025
Thiamin HCl	1.0	0.05		10
Pyridoxin HCl	0.5	0.25		1.0
Nicotinic acid	0.5	0.25		1.0
Glycine	2.0	1.0 ⎭		-
Glutamine	-	800 ⎫	for pollen	800
1-serine	-	100 ⎬	culture	-
m-inositol	-	5000 ⎭	(Nitsch, 1974)	100
Sucrose	90 000 to 120 000	20 000		100 000
Kinetin	0.5	-		-
2,4-D	2.0	-		0.1
NAA	-	-		0.1

pH adjusted to 5.8 before medium is autoclaved.
Agar used at 7000 to 10 000, activated charcoal at 3000 to 10 000.

For pollen culture (tobacco) precultured anthers are gently crushed in liquid medium to give a pollen density of about $4-5 \times 10^4$ grains per ml, which is equivalent to the contents of one anther per ml. The suspension is filtered through nylon mesh (50–100 µm) and the filtrate centrifuged at 100 g for about 5 min. The supernatant is discarded. The pollen, after being washed at least once with liquid medium, is resuspended in the liquid medium at the initial density and transferred to Petri dishes. The dishes are kept stationary.

Recommended media are given in Tables 1 and 2. Potato media (Table 1) are invaluable in cereal anther culture. In the Solanaceae hormones should be used sparingly, if at all, because of their stimulating effect on somatic tissues.

Incubation of cultures

Temperature   Normally 25–30°C is required. Still higher temperatures are used in brassica anther culture (Keller & Stringham, 1978).

Light   This has not been critically examined. In general, cultures are incubated in darkness and transferred to light when the embryos or callus have formed.

Regeneration of plantlets from callus

Hormones are usually omitted from the medium and the sugar level reduced to 2 or 3% (Table 1). The auxin and mineral salt composition may be changed. Lower incubation temperatures are recommended, e.g. 15–20°C.

Maintenance of plantlets before transfer to soil

Plantlets need to be transplanted individually in fresh medium and kept in light until a good root system is established. The mineral salt composition and the sugar content of the medium can be further reduced (Table 1).

Transfer of plantlets to soil

Well-rooted plantlets should be removed from the medium with as little damage to root tips as possible. The complete removal of agar is not essential. An initial transfer to sterile soil or to mixtures of sterile peat and Perlite is recommended, as is maintenance in mist until the plantlets are established.

CONCLUSION

There are marked genotypic differences in the response of anthers both between and within species. Protocols for legumes, palms and other economically important plants have not yet been devised. Albinism is a major drawback in grasses. Selective procedures favouring the develop- ment of green plants need intensive study. In pollen culture, methods of preserving pollen viability both during and after isolation need to be devised.

REFERENCES

References marked with an asterisk are recommended for general reading.

Carlson P.S. (1973) Methionine-sulfoximine-resistant mutants of tobacco.
  Science 180, 1366-8.
Chu C-C. (1978) The N$_6$ medium and its application to anther culture of
  cereal crops. *Proceedings of Symposium on Plant Tissue Culture*, pp.
  43-50. Science Press, Peking.
Chuang C-C., Ouyang T-W., Chia H., Chou S-M. & Ching C-K. (1978) A set
  of potato media for wheat anther culture. *Proceedings of Symposium
  on Plant Tissue Culture*, pp. 51-65. Science Press, Peking.
Dunwell J.M. (1979) Anther culture in *Nicotiana tabacum*: the role of
  the culture vessel atmosphere in pollen embryo induction and growth.
  Journal of Experimental Botany 30, 419-28.
Guha S. & Maheshwari S.C. (1964) *In vitro* production of embryos from
  anthers of *Datura*. *Nature* 204, 497.
Hu H. (1978) Advances in anther culture investigations in China.
  Proceedings of Symposium on Plant Tissue Culture, pp. 3-10. Science
  Press, Peking.
Jensen C.J. (1974) Chromosome doubling techniques in haploids.
  Haploids in Higher Plants: Advances and Potential (Ed. by K.J. Kasha),
  pp. 153-90. University Press, Guelph.
*Keller W.A. & Stringham G. (1978) Production and utilization of
  microspore-derived haploid plants. *Frontiers of Plant Tissue Culture,
  1978* (Ed. by T.A. Thorpe), pp. 113-22. International Association for
  Plant Tissue Culture, Calgary.
Maia E., Bettachini B. & Maia N. (1975) Obtention de plantes exemptes
  de virus par cultière *in vitro* d'anthères de tabac infectés par le
  virus de la Mosaique du Tabac. *Comptes rendus hebdomadaire des
  séances de l'Académie des sciences, Paris* 280, 2505-8.
Miao S-H., Kuo C-S., Kwei Y-L., Sun A-T., Ku S-Y., Lu W-L., Wang Y-Y.,
  Chen M-L., Wu M-K. & Hang L. (1978) Induction of pollen plants of
  maize and observations on their progeny. *Proceedings of Symposium on
  Plant Tissue Culture*, pp. 23-33. Science Press, Peking.
Mix G., Wilson H.M. & Foroughi-Wehr B. (1978) The cytological status of
  plants of *Hordeum vulgare* L. regenerated from microspore callus.
  Zeitschrift für Pflanzenzüchtung 80, 89-99.
Nitsch C. (1974) Pollen culture – a new technique for mass production
  of haploid and homozygous plants. *Haploids in Higher Plants:
  Advances and Potential* (Ed. by K.J. Kasha), pp. 123-35. University
  Press, Guelph.
*Nitszche W. & Wenzel G (1977) Haploids in plant breeding. *Fort-
  schritte der Pflanzenzüchtung* Supplement No. 8, pp. 1-80.
Sopory S.K., Jacobsen E. & Wenzel G. (1978) Production of monohaploid
  embryoids and plantlets in cultured anthers of *Solanum tuberosum*.
  Plant Science Letters 12, 47-54.
Sunderland N. (1978) Strategies in the improvement of yields in anther
  culture. *Proceedings of Symposium on Plant Tissue Culture*, pp. 65-
  86. Science Press, Peking.
*Sunderland N. & Dunwell J.M. (1977) Anther and pollen culture. *Plant
  Tissue and Cell Culture* (Ed. by H.E. Street) 2nd edition, pp. 223-65.

Blackwell Scientific Publications, Oxford.

Sunderland N. & Roberts M. (1979) Cold-pretreatment of excised flower buds in float culture of tobacco anthers. *Annals of Botany* 43, 405-14.

Sunderland N. & Wildon D.C. (1979) A note on the pretreatment of excised flower buds in float culture of *Hyoscyamus* anthers. *Plant Science Letters* 15, 169-75.

Wenzel G., Hoffmann F. & Thomas E. (1977) Increased induction and chromosome doubling of androgenetic haploid rye. *Theoretical and Applied Genetics* 51, 81-6.

Wilson M.H. (1977) Culture of whole barley spikes stimulates high frequencies of pollen cultures in individual anthers. *Plant Science Letters* 9, 233-8.

Cereal tissue culture

E. THOMAS*, R. BRETTELL†, I. POTRYKUS† &
W. WERNICKE†
*Rothamsted Experimental Station, Harpenden,
Hertfordshire AL5 2JQ
†Friedrich Miescher-Institut, PO Box 273,
CH-4002 Basel, Switzerland

INTRODUCTION

The availability of large numbers of haploid and/or diploid plant cells,
protoplasts and microspores from which plantlets can be regenerated, and
of techniques for single cell cloning, genetic manipulation and cell
selection could make possible many advances in the cell physiology and
genetics of higher plants comparable to those already achieved with
micro-organisms. Further, genetically modified plants, if properly
utilized, show promise of providing breeders with beneficial gene
combinations (Thomas et al., 1979). Apart from the aims of hybridization through protoplast fusion of sexually incompatible species and of
modification by transfer of organelles or nucleic acids, there are more
immediate goals. For example, it should be possible to subject a
population of single mutated cereal cells to the action of a fungal
or bacterial toxin (Brettell et al.; Strauss et al.; this volume),
to select cell clones resistant to the toxin and eventually to
regenerate resistant plants which can be incorporated into conventional
breeding programmes or used to study the nature of resistance. More
detailed articles on the uses and potential uses of tissue culture in
plant pathology and plant breeding are provided by Ingram and by Day
(this volume). Tissue culture techniques, especially at the single cell
and protoplast level, are relatively new and there are still numerous
obstacles to be overcome before they are fully applicable to cereals
(King et al., 1978). The type of model cereal which could prove
extremely useful for carrying out in vitro genetics would be able to
undergo all or most of the manipulations depicted in Fig. 1.

It is not the aim of this communication to indicate how tissue
cultures of cereals can be used for solving some of the problems facing
plant pathologists, nor to assess the value of the cultures. Rather the
aim is to point out which tissue culture systems are available, how they
may be established and how they compare with the model depicted in Fig.
1.

Ingram D.S. & Helgeson J.P. (1980) *Tissue Culture Methods for Plant Pathologists*

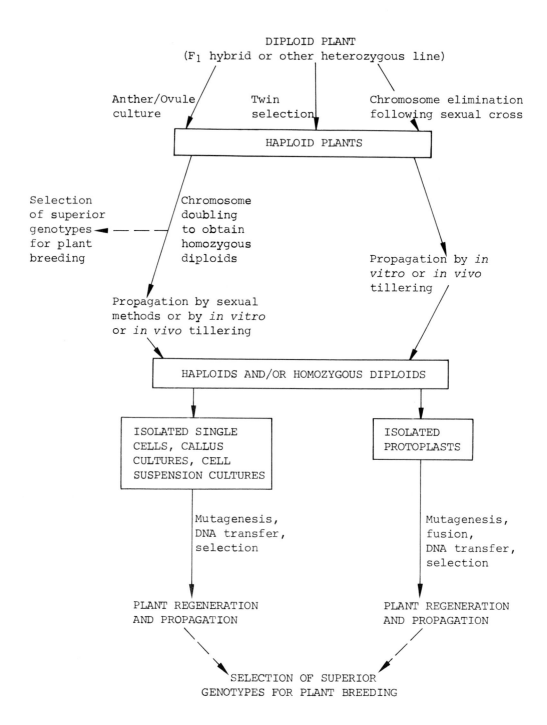

Figure 1. Procedures used in the *in vitro* genetic manipulation of cereals.

HAPLOID PLANTS AND THEIR PRODUCTION

The apparent inadequacy of multicellular diploid material for mutation research with higher plants has been documented by Devreux & de Nettancourt (1974). It seems that the solution to these inadequacies is to use haploid tissues and to induce and screen mutants at the single cell level. Theoretically in cereals, even without mutation and single cell or protoplast culture, the production of haploids through the culture of microspores or macrospores, if properly utilized, should contribute to plant breeding. Melchers (1972) pointed out that the value of microspore culture 'lies in the speed with which homozygous lines can be obtained compared with conventional techniques of inbreeding; in many plants which are self sterile, haploid production followed by diploidisation may be the only way of achieving homozygosity. The objection that today one does not strive to obtain homozygous material because heterozygotes may be more vigorous, is far too short-sighted to be taken in earnest by plant breeders; the possibility of obtaining exactly reproducible seeds again and again from (by crossing) truly homozygous starting lines is rather an interesting advantage for the breeder.' Another advantage of microspore culture would be to assist breeding programmes that require a combination of dominant alleles (e.g. in programmes searching for disease resistance). Further, when large numbers of different microspores can be successfully cultured from F_1 hybrids there is the possibility of obtaining, in one step, plantlets which are of better quality than either of the parents used in F_1 production. Using microspore-derived plantlets several laboratories have reported obtaining superior tobacco lines with increased disease resistance (Nitzsche & Wenzel, 1977).

There are several methods for producing haploids in cereals. Over the last 10-12 years the technique of anther culture has been widely used in attempts to mass produce haploids from microspores (immature pollen grains) and even though numerous cereal species are responsive in anther culture (Nitzsche & Wenzel, 1977; Sunderland, this volume) there are several problems limiting the use of this technique as a source of homozygous plants for use in crop improvement. The yield of plantlets per plated anther is generally very low, varies widely and is strongly dependent upon the physiological condition of the donor plant. The yield of potentially useful plants is often further reduced by the production of large numbers of albino plantlets and by the formation of heterozygous plantlets from unreduced microspores (those which have not passed fully through meiosis).

The techniques of anther culture in cereals are relatively simple and involve transferring intact sterile anthers at the correct developmental stage to a culture medium which generally, but not always, contains 2,4-dichlorophenoxyacetic acid (2,4-D). However, it is not always easy to obtain reproducible results throughout the year. For this reason we shall describe a successful method of standardizing Secale cereale donor material by the pretreatment of excised flower spikes (Thomas et al., 1975). Using this technique plantlets can be obtained from microspores throughout the year.

Following vernalization, seedlings should be grown to flowering in glasshouses maintained at 20°C for a 16 h light period and at 18°C for an 8 h dark period. During the winter months normal day length should be artificially extended to 16 h by Philips HQL 400 W/R lamps or by BLV 750 W halogen lamps (or equivalent) providing light of approximately 6000 lx measured at the top of the plants. Our evidence indicates that microspores cultured at about the period of the first microspore mitosis are most likely to undergo embryogenesis. Although this microspore stage is present in anthers of some unemerged spikes, the spikes usually contain this stage after they have broken through the enclosing flag leaves.

In order to obtain microspores during mitosis and to prevent microbial contamination of emergent spikes the following procedure should be adopted. Harvest the spikes while they are still enclosed within their flag leaves, and surface sterilize with 70% alcohol soaked in cotton wool. Immerse the cut ends of the spikes in the liquid medium of Nitsch & Nitsch (1969) lacking phytohormones and contained in 100 or 250 ml Erlenmeyer flasks. Cover the spikes and the flask necks with sterile plastic bags, seal tightly with adhesive tape and incubate at 26°C under a 12 h light cycle of 7000 lx. The spikes develop further and contaminant-free anthers containing microspores in the required developmental stage will be obtained. Developmental stages should be monitored by squashing a single anther from each floret in acetocarmine.

It has been found that periods of cold treatment improve the response of microspores in culture. The flasks and the emerged spikes should therefore be transferred to a dark refrigerator at 6°C for 6-10 days. Although almost any basal nutrient culture medium containing 2,4-D (1-2 mg/l) is suitable for obtaining plantlet production from excised anthers, the best results are obtained using the culture medium containing potato extract which was developed for rice by the Institute of Genetics, Peking. Using this medium embryos, plantlets and calluses arise from the cultured anthers after 6-8 weeks; at this stage it may be necessary to transfer them to another culture medium to encourage further growth. Full details of these media are given by Wenzel et al. (1977).

The difficulty in obtaining high yields and the need for a large labour force severely limit the widespread use of anther culture in cereals. A very successful alternative method for the production of haploid plants of certain species involves conventional methods coupled with tissue culture techniques. When *Hordeum vulgare* is crossed with *Hordeum bulbosum* fertilization occurs in the usual manner. There is, however, a subsequent elimination of *H. bulbosum* chromosomes resulting in the production of haploid embryos containing only *H. vulgare* chromosomes. The endosperm of the seed fails to develop beyond an early stage and consequently it is essential to isolate the embryos and to nurture them carefully *in vitro*. Technical procedures for achieving this as well as source literature are reviewed by Jensen (1977). Chromosome elimination using *H. bulbosum* as the pollen parent has also been used for haploid production in other cereal genera, e.g. *Triticum*. San Noeum (1976) and Jensen (personal communication) have also reliably produced haploids by the culture of unfertilized *Hordeum* ovules.

Cereal plants which have a tendency to form tillers under field conditions also do so under greenhouse or *in vitro* conditions, and for these plants (e.g. *Sorghum*, *Hordeum*) it is generally not difficult to propagate haploids vegetatively once they are obtained. For plants such as *Zea mays* vegetative propagation is much more difficult and great care must be taken not to kill available haploid plants during attempts to diploidize them. Diploidization of most cereals can be achieved by treating young haploid plantlets with colchicine as described by Jensen (1977) for barley.

It remains to be shown whether the use of cereal homodiploids is advantageous in plant breeding programmes. Results obtained with rice (Yin *et al.*, 1976) and barley (Reinbergs *et al.*, 1976) have demonstrated that homozygous lines produced using tissue culture techniques lead to material comparable to that produced by conventional breeding techniques, but more rapidly.

CALLUS AND CELL SUSPENSIONS

Prior to the discovery that plant tissues could be enzymatically broken down into isolated single cells and/or protoplasts, the general approach to callus and cell suspension culture of dicotyledonous species was to excise a piece of plant tissue (e.g. root, leaf, flower-bud etc.) under aseptic conditions and to transfer this to a culture medium solidified with agar and containing auxins and/or cytokinins. Under suitable conditions tissue treated in this way will undergo cell divisions giving rise to a largely unorganized mass of proliferating tissue commonly called a callus. Such calluses can be propagated indefinitely by regular subculture to fresh nutrient medium. If the callus culture is friable enough it is often possible to transfer it to a liquid culture medium and, by shaking on a horizontal rotary shaker, to establish a suspension culture consisting of single cells and small cell aggregates. Through filtration and plating it may even be possible to establish cell clones of single cell origin from which plantlets can be regenerated. Such cultures may be of use in the production of mutant lines.

In cereals the establishment of unorganized callus and cell suspension cultures is a comparatively rare event (King *et al.*, 1978). Usually the cultures consist of root or shoot primordia interspersed with large, highly vacuolated, terminally differentiated cells. Most calluses and cell suspensions so far obtained in cereals are non-morphogenic and of little use for applied purposes. On a more optimistic note it should be added that Vasil & Vasil at the University of Florida (unpublished data) have reported obtaining suspension cultures of pearl millet, *Pennisetum americanum*, which can be plated onto a medium solidified with agar where they give rise to calluses from which plantlets arise. It will be interesting to see whether these techniques are also applicable to other species. Because of the difficulties in obtaining true callus and/or true suspension cultures we prefer to

describe the cultures which we have obtained in our studies as organized
cultures. The establishment of cultures of this type is described in
the next section.

ORGANIZED TISSUE CULTURES

Root-forming systems

Most cereal cultures produce roots. The experimental procedure used
to establish root-forming cultures from *S. cereale* haploid plants is
described below. Such rooting cultures may have some applied uses:
for example, as a source of material for maintaining or propagating
nematodes (Krusberg & Babineau, 1979). Under certain conditions they
can also give rise to plantlets (see following section); therefore, if
rooting cultures are obtained following a procedure of genetic manipulation, it is advisable not to discard them immediately.

    In our experiments root-forming cultures were obtained from twin
selected haploid plants of *S. cereale* cv. Somro (provided by J. Straub,
Max-Planck Institut, Cologne-Vogelsang). The methods described here may
be used with similar material. Plants should be propagated through
tillers in a greenhouse at 12-15°C under short days (8 h light, 16 h
dark) and induced to flower by transference to 22-25°C under long days
(16 h light, 8 h dark). Flowering stems should be swabbed with 70%
alcohol and the nodes of the stems excised and sterilized by transfer to
a 5% solution of sodium hypochlorite for 10 min. After thorough washing
in several changes of distilled water the ends of the nodes should be
cut off to expose the living tissue. The nodes should then be transferred to the agar medium of Nitsch & Nitsch (1969) with the addition of
2,4-D (1 mg/l). Following incubation in the dark at 25°C for 2-3 weeks
the node cultures undergo proliferation and spontaneously produce roots.
These rooting cultures can be maintained by subculture to fresh medium
at 4- to 6-week intervals.

Plant forming systems

Plants from roots   Shoot formation can be obtained in *S. cereale* by
transfer of the root-forming cultures of flower-node origin described
above to the agar medium of Murashige & Skoog (1962) with the addition
of 1-naphthaleneacetic acid (NAA), 6-benzylaminopurine (BAP) and
gibberellic acid, each at 2 mg/l. In 20% of the cultures the roots of
the cultures produce proliferations at those ends in contact with the
new culture medium, and these spontaneously produce green plantlets
(*S. cereale* is quite unusual in being able to produce shoots from roots).
Shoots can be rooted by transfer to Murashige & Skoog medium lacking
phytohormones, and once the plantlets have established good root systems
they should be washed free of agar with tap water, transferred to
potting compost and covered with an inverted beaker to minimize
desiccation. After about 7 days the beakers may be removed.

Plants from cultures of immature embryo origin   This method is the

most widely applicable in cereals and has been described in detail by
Green & Phillips (1975) for immature embryos of *Z. mays*. We have
applied the technique to *Sorghum bicolor* (Thomas *et al.*, 1977) and
there would appear to be no reason why it should not be applicable to
all cereal species. The technique is as follows.

Remove seeds from glasshouse-grown plants approximately 10-30 days
after pollination (just before the seed breaks through the enclosing
bracts), surface sterilize (0.01% $HgCl_2$ + 1 drop Tween 80/100 ml) and
wash six times in sterile water. Isolate immature embryos approximately
1 mm in length from the seeds and place on the agar medium of Murashige
& Skoog containing sucrose (3%) and 2,4-D (2.5 or 5.0 mg/l) with the
scutellar side uppermost and the coleorhiza-coleoptile side in contact
with the medium. Incubate under diffuse light of 1000 lx at 27°C (16 h
light/8 h dark). After only 15 days of incubation many hundreds of
secondary embryos and plantlets arise on the surface of the proliferating scutellum. Cultures capable of continuous shoot production can be
obtained by transferring the proliferating cultures to fresh medium
containing BAP (0.5 mg/l) and either 2,4-D (2 mg/l) or NAA (5 mg/l;
Dunstan *et al.*, 1978). A medium lacking phytohormones but containing
100 mg/l adenine sulphate encourages vigorous plantlet formation; these
plantlets form roots spontaneously and can be potted as described
previously for rye.

The immature embryo system could be valuable for producing haploid
shoot-forming cultures in barley. Haploid immature embryos would have
to be produced using the *H. vulgare-H. bulbosum* chromosome elimination
method; then, instead of the embryos being grown on to plantlets, they
could be proliferated. Similar experiments have been carried out by
Dhaliwal & King (unpublished data) using haploids of *Z. mays* produced
from different genetic stocks. These systems are described in more
detail by King *et al.* (1978). The great value of such shoot-forming
haploid cultures would probably be as a source of protoplasts.

Plants from unemerged inflorescences   This method of obtaining shoot-
forming cultures is relatively simple and involves the transfer of young
inflorescence segments to media similar to those used for other cereal
tissue cultures. A reproducible method is currently being worked out
for *S. bicolor* (Brettell, Wernicke & Thomas, unpublished data). The
method has several advantages. It can be used for the vegetative
propagation of haploids identified under field conditions and for the
propagation of heterozygous diploid species which show some interesting
character which is likely to be masked if sexual methods are used. The
method is less exacting than careful crossing and monitoring of
pollinated plants and is also applicable to some other cereals such as
maize (Gordon, personal communication) and wheat (Dudits *et al.*, 1975).

PROTOPLAST CULTURES

In dicotyledonous species the most successful methods of protoplast
culture involve mesophyll tissue. Unfortunately, despite extensive

efforts, attempts to induce division of cereal leaf mesophyll protoplasts have largely failed. The only convincing report of sustained
division of protoplasts isolated direct from intact cereal plants is
that of Potrykus *et al.* (1977) who used stem protoplasts of corn. This
result has not been repeated despite many attempts, nor did the cultures
give rise to plants (Potrykus & Thomas, unpublished data). It may be
easier to grow protoplasts from rapidly dividing callus cultures or
suspension cultures. The report of Vasil & Vasil that morphogenic
suspension cultures of *P. americanum* may be routinely obtained, is
encouraging, especially since they were able to isolate protoplasts from
the suspensions and induce them to divide and initiate plantlets from
the resulting cell colonies (Vasil & Vasil, 1979). It is to be hoped
that in the near future techniques similar to those reported for
P. americanum will be applicable to other cereals.

REFERENCES

Devreux M. & de Nettancourt D. (1974) Screening mutations in haploid
    plants. *Haploids in Higher Plants - Advances and Potential* (Ed. by
    K.J. Kasha), pp. 309-22. University of Guelph, Ontario, Canada.
Dudits D., Nemet G. & Haydu Z. (1975) Study of callus growth and organ
    formation in wheat (*Triticum aestivum*) tissue cultures. *Canadian
    Journal of Botany* 53, 957-63.
Dunstan D.I., Short K.C. & Thomas E. (1978) The anatomy of secondary
    morphogenesis in cultured scutellum tissues of *Sorghum bicolor*.
    Protoplasma 97, 251-60.
Green C.E. & Phillips R.L. (1975) Plant regeneration from tissue
    cultures of maize. *Crop Science* 15, 417-21.
Jensen C.J. (1977) Monoploid production by chromosome elimination.
    Plant Cell, Tissue and Organ Culture (Ed. by J. Reinert &
    Y.P.S. Bajaj), pp. 299-340. Springer-Verlag, Berlin.
King P.J., Potrykus I. & Thomas E. (1978) *In vitro* genetics of cereals:
    problems and perspectives. *Physiologie Végétale* 16(3), 381-99.
Krusberg L.R. & Babineau D.E. (1979) Application of plant tissue
    culture to plant nematology. *Plant Cell and Tissue Culture,
    Principles and Applications* (Ed. by W.R. Sharp, P.O. Larsen,
    E.F. Paddock & V. Raghavan), pp. 401-19. Ohio State University Press,
    Columbus.
Melchers G. (1972) Haploid higher plants for plant breeding. *Zeit-
    schrift für Pflanzenzüchtung* 67, 19-32.
Murashige T. & Skoog F. (1962) A revised medium for rapid growth and
    bioassays with tobacco tissue cultures. *Physiologia Plantarum* 15,
    473-97.
Nitsch J.P. & Nitsch C. (1969) Haploid plants from pollen grains.
    Science 163, 85-7.
Nitzsche W. & Wenzel G. (1977) Haploids in Plant Breeding. *Advances in
    Plant Breeding, Zeitschrift für Pflanzenzüchtung*, Supplement 8.
    Verlag Paul Parey.
Potrykus I., Harms C.T., Lörz H. & Thomas E. (1977) Callus formation
    from stem protoplasts of corn (*Zea mays* L.). *Molecular and General
    Genetics* 156, 347-50.

Reinbergs E., Park S.J. & Song L.S.P. (1976) Early identification of superior barley crosses by the doubled haploid technique. *Zeitschrift für Pflanzenzüchtung* 76, 215-24.

San Noeum L.H. (1976) Haploides d'*Hordeum vulgare* par culture *in vitro* d'ovaires non fécondes. *Annales de L'Amélioration des Plantes* 26, 751-4.

Thomas E., Hoffmann F. & Wenzel G. (1975) Haploid plantlets from microspores of rye. *Zeitschrift für Pflanzenzüchtung* 75, 106-13.

Thomas E., King P.J. & Potrykus I. (1977) Shoot and embryo-like structure formation from cultured tissues of *Sorghum bicolor*. *Naturwissenschaften* 64, 587.

Thomas E., King P.J. & Potrykus I. (1979) Improvement of crop plants via single cells *in vitro* - an assessment. *Zeitschrift für Pflanzenzüchtung* 82, 1-30.

Vasil V. & Vasil I.K. (1979) Isolation and culture of cereal protoplasts II. Embryogenesis and plantlet formation from protoplasts of *Pennisetum americanum*. *Theoretical and Applied Genetics* 56, number 3, 97-100.

Wenzel G., Hoffmann F. & Thomas E. (1977) Increased induction and chromosome doubling of androgenetic haploid rye. *Theoretical and Applied Genetics* 51, 81-6.

Yin K-Ch., Hsü Ch., Chu Ch.-Y., Pi F.-Y., Wang S.-T., Liu T.-Y., Chu Ch.-Ch., Wang Ch.-Ch. & Sun Ch.-S. (1976) A study of the new cultivar of rice raised by haploid breeding method. *Scientia Sinica* 19, 227-42.

In vitro propagation

G. HUSSEY

John Innes Institute, Colney Lane,
Norwich NR4 7UH

INTRODUCTION

Efficient vegetative propagation is essential for the breeding and
exploitation of most heterozygous crops. Tissue culture methods have
greatly increased the scope and potential of propagation by exploiting
regenerative behaviour more efficiently and in a wider range of plants
than is possible with conventional procedures. Disease-free material
(often obtained with difficulty) is eminently suited to *in vitro*
techniques on account of the built-in disease protection it provides.
Cloned individual plants of heterozygous species generally propagated
by seed may be more useful in experimental work because of much
greater uniformity.

This paper presents general guidelines for methods of plant
propagation by tissue culture with special reference to the question
of genetic stability.

Methods of in vitro *propagation and genetic stability*

All vegetatively propagated plants are susceptible to occasional
mutation. During clonal increase it is essential that tissue culture
in itself should not significantly increase the mutation rate. The
genetic uniformity of the propagules and their susceptibility to
mutation *in vitro,* however, depends both on the nature of the starting
material and on the methods subsequently used for multiplication
(Hussey, 1978).

Vegetative propagation consists of the formation and multiplication
of shoot meristems, each meristem being capable of forming an entire
plant. Tissue cultures for propagation may be started from existing
meristems (embryo, main shoot or axillary shoots; Fig. 1,a) or from
organ explants suitable for the induction of adventitious meristems
(shoot apices or embryos), either directly from organ tissue or via
intermediate callus or suspension cultures (Fig. 1,b).

Ingram D.S. & Helgeson J.P. (1980) *Tissue Culture Methods for Plant Pathologists*

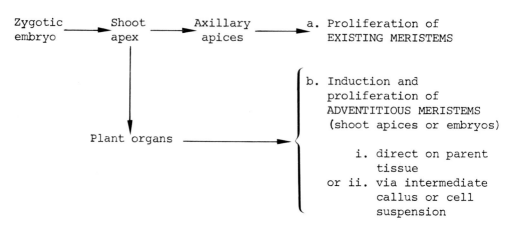

Starting material   Tissue culture
from conventionally grown plant

Zygotic ⟶ Shoot ⟶ Axillary ⟶ a. Proliferation of
embryo   apex   apices   EXISTING MERISTEMS

                                          b. Induction and
                                             proliferation of
                                             ADVENTITIOUS MERISTEMS
                                             (shoot apices or embryos)

Plant organs ⟶

                                             i. direct on parent
                                                tissue
                                          or ii. via intermediate
                                                 callus or cell
                                                 suspension

Figure 1.   Scheme for *in vitro* propagation.

Multicellular organized shoot meristems are the most genetically
stable parts of the plant, spontaneous or artifically induced mutant
cells usually being eliminated by diplontic selection or by the
formation of unstable chimeras. Adventitious meristems, by contrast,
can give rise to wholly mutant propagules as they are usually derived
from single or very small groups of cells. Although the occurrence
of spontaneous mutations may possibly be enhanced as a result of the
in vitro conditions themselves, the kind of organ used for the induction
of adventitious shoots seems to be a more important factor in the
production of mutant propagules. In most plant species the process of
differentiation involves the formation of cells with permanently
modified genomes (e.g. polyploidy, somatic crossing over, etc.) and
plants derived from such cells may differ significantly from the parent
type (Skirvin, 1978).

In organ explants the formation of adventitious meristems begins
with the stimulation of cell division by auxin or cytokinin or the two
together; these divisions may give rise to meristematic structures
either within the explant tissue itself or after varying amounts of
intermediate callus growth, according to the hormones used and their
concentrations. For the production of adventitious shoots as free as
possible from any genetic changes the cultures should begin with the
youngest, least differentiated tissue available, to minimize the
involvement of aberrant cells.

The number of adventitious shoots obtainable direct on organ
explants will be limited by the size of the explants, but the formation
of callus can provide extra tissue for the induction of more shoots.

Callus, however, especially the fast growing tissue induced with higher hormone concentrations, is liable to produce polyploid and other abnormal cells which may be preferentially selected by the nutrient medium. Serially propagated callus typically accumulates a high proportion of polyploid, aneuploid and other mutant cells and may eventually lose the capacity to regenerate shoots. Callus induced from mature organs already containing cells with modified genomes will develop these properties more quickly.

GENERAL PROCEDURES

The general sequence of procedures is: selection of a mother plant; choice of an explant; sterilization and washing of the explant; transfer of the explant to establishment medium; proliferation of shoots on multiplication medium; transfer of shoots to a rooting medium (or a storage medium); and planting out of propagules. These procedures, together with appropriate general information on media and controlled environment, are dealt with in this and the following two sections. These descriptions provide guidelines only, and modifications to suit individual species should be adapted from information in papers dealing with similar types of plants. More general discussion of *in vitro* propagation is provided by Murashige (1974), Debergh (1977), Reinert & Bajaj (1977) and Thorpe (1978).

Selection of mother plant(s)

Plants should be as healthy as possible and at the beginning of active growth. Dormant or semi-dormant plants may not react well. Bulbs, corms, tubers and other plants requiring a period of high or low temperature to break dormancy should have had the necessary temperature treatment to encourage sprouting.

Choice of explant

To ensure the maximum genetic stability cultures should be started with embryos, shoot apices or meristematic organ tissue.

Seed may be germinated aseptically *in vitro* or, alternatively, the embryos excised before germination. Main shoot apices can be dissected from sprouted seeds, larger seedlings, or the shoots of mature plants. The internodes of plants with stems will contain axillary shoots in various stages of development. Axillary shoots from rosette plants are best obtained by removing the leaves in turn, causing as little damage as possible to the proximal ends where the buds are present. Bulbs and corms contain variable numbers of axillary buds according to the species. Rhizomatous plants have their vegetative meristems in the subterranean organ; the rhizome tip, although difficult to sterilize, is usually well protected by covering structures.

In plants where axillary meristems are not readily formed

adventitious meristems can be induced on explants from any of the main
organs that can be obtained aseptically or satisfactorily decontaminated.
In dicotyledons young leaves dissected from actively growing shoots are
a good source of meristematic tissue. In monocotyledons the meristems
of leaves and scales are located at the proximal end where they join
the basal plate. Leaf-base and scale-base explants should include a
small piece of basal plate tissue.

Sterilization and washing of the starting material

Whenever possible clean plants should be grown in an uncontaminated
growing medium such as sterilized Perlite. Overhead watering and
splashing of foliage should be avoided.

Shoot or meristem tips with a sufficient covering of mature leaves
or scales, or the centres of bulbs, corms or rhizomes, should be
dissected out sterile by gently peeling off the outer structures after
wiping them with 70% ethanol. Alcohol or other sterilizing fluids
should be used sparingly with loose buds to prevent penetration by
capillary action.

Organ explants should be sterilized in *freshly* diluted (1 in 20)
sodium hypochlorite solution (1% w/v available chlorine) containing
0.1% Tween 20 for 5-30 min according to the thickness of the cuticle.
The aim is to kill surface micro-organisms with minimal damage to the
outer cell layers. The younger the organ the less it will probably
be contaminated, but the more sensitive it will be to sterilants. Two
rinses in sterile water will remove the hypochlorite.

Where aseptic shoot apices or young vegetative organs are required
an alternative technique is to raise sterile seedlings in a suitably
transparent and sterilizable container such as a glass casserole dish
containing Perlite or Vermiculite.

Transfer of explants to medium

Explants are usually transferred to agar medium, although some plants
culture well only in liquid medium (see below). When using solid
medium, horizontally placed explants should be pressed firmly against
the agar surface to ensure good contact; vertically placed explants
may have up to one-third immersed in the medium.

Vessels suitable for culturing initial explants are 25 ml screw-
topped universal glass tubes or 50 ml Erlenmeyer flasks closed with
aluminium foil. Proliferating shoots may be cultured in larger flasks
or 100-300 ml screw-topped pre-sterilized polystyrene jars. For larger
scale shoot production various glass or plastic containers may be used
according to the size and habit of the propagules. Culture vessels
should contain a layer of medium 10-15 mm deep.

The factors listed below may affect the reactions of the explant
and should be empirically tested where possible, particularly if poor

results are obtained initially.
  1. The orientation of the explant. 90° or 180° to the normal
     morphological orientation may improve or inhibit regeneration.
  2. The degree of immersion in the medium. The more the explant is
     immersed the less oxygen it will get.
  3. The type of covering, affecting the gaseous exchange within and
     without the vessel.
  4. The ratio of explant bulk to quantity of medium. Several explants
     in the same tube may do better than only one.

Nutrient media

There are a number of standard salt mixtures (Gamborg *et al.*, 1976),
the most widely used being that of Murashige & Skoog (1962) which may
be used at full strength or at half or less dilution. Appropriate stock
solutions are given by Helgeson (this volume).

Vitamins and hormones may be made up in sterile water as 100 mg/l
stock solutions and kept in the dark at 5°C (except 6-benzylaminopurine
(BAP) which should be kept in the dark at room temperature).

Vitamins should be used at the following concentrations (mg/l):
thiamine hydrochloride (0.5); pyridoxine hydrochloride (1.0); nicotinic
acid (5.0). Myo-inositol may be added as a general supplement at 100
mg/l. Adenine sulphate at 150 mg/l may be beneficial for shoot
proliferation. With many species growth is assisted by casein hydro-
lysate at 500-1000 mg/l.

The control of organogenesis in cultured plant tissues derives
largely from the work of Skoog & Miller (1957). According to the
Skoog-Miller model the formation of shoots is promoted by high levels
of cytokinin relative to auxin, whereas the reverse favours the
development of roots. Three cytokinin compounds are currently in
general use for propagation: kinetin, BAP and 6-(γ,γ,dimethylallyl-
amino)-purine (2iP), with kinetin generally being the least and 2iP
the most active. BAP is the most useful and reliable compound and
should be tried first.

Many different auxins are available. The naturally occurring auxin
indole 3-ylacetic acid (IAA) is the least active and stable, and the
stronger and more stable synthetic analogues such as indole-3-butyric
acid (IBA) and 1-naphthaleneacetic acid (NAA) are more often used.
The highly active compound 2,4-dichlorophenoxyacetic acid (2,4-D) is
the most effective synthetic auxin for promoting callus but strongly
inhibits shoot formation. These cytokinins and auxins are active over
a wide range of concentrations, from 0.01-30 mg/l, but the range 0.1-10
mg/l should be tried first.

Medium is generally solidified with 0.7-0.8% agar after adjusting
the pH to 5.6-5.8. The gel must be firm enough to support explants
or shoot clusters but if it is too hard it may prevent adequate contact
between the medium and the tissues, especially after some drying out.

Some plants culture well only in agitated liquid medium, while an alternative method utilizes a filter paper bridge or wick to support the explant above the liquid, with the free ends of the paper immersed.

Controlled environment

Except for a short period just before planting out, little or no photosynthesis need take place in cultured organs and shoots; only low light intensities in the order of 1000-5000 lx are therefore required. Banks of fluorescent tubes with a 16 h daylength are widely used, although a few reports suggest that both intensity and daylength may be critical for shoot formation in certain species. Other appropriate daylengths should be tried if there is any known photoperiodic effect on vegetative growth.

It should be noted that in containers with opaque screw-top lids or aluminium foil closures the light intensities will be 20-60% lower (according to spacing) than in completely transparent vessels.

Most cultures will grow within the range 15-25°C. The normal environmental requirements of each species should be taken into account when deciding on the particular temperature to be used.

Cultures should normally be illuminated from above, but with most plants fluorescent tubes may be placed below or alongside the containers without ill effects. Whatever the controlled environment installation used, it is important that the individual containers are either in an even temperature or are warmer towards the top. If they are cooler towards the top condensation may block the narrow spaces between the containers and the covering, especially if there are any diurnal temperature changes, allowing the entry of micro-organisms. A suitable temperature gradient is most easily achieved by having most of the lighting from above and the containers supported on an open structure such as wire mesh that will allow air to circulate freely.

METHODS OF SHOOT INDUCTION AND PROLIFERATION

Axillary shoots

In seedlings of most higher plants the main shoot apex produces a succession of leaves, each of which carries in its axil a replica of the main apex capable in turn of producing its own axillary apices. This system has obvious potential for exponential increase in numbers of meristems, but such an increase can be realized only by the relaxation of the endogenous hormonal constraints which normally hold the system in check and control the degree of branching of the plant. This relaxation can be achieved in culture to varying extents by the inclusion in the medium of cytokinin, which stimulates premature development and outgrowth of axillary meristems. Shoots cultured on cytokinin characteristically develop into miniature clusters rather than single axes. Once a cluster has developed to the stage where it

can be split up into separate shoots or smaller clumps, the resulting
pieces may be subcultured to produce further shoot clusters. If the
basic medium is adequate to sustain normal shoot growth this process
may apparently be continued indefinitely without any physiological
deterioration taking place.

The maximum rate at which axillary shoots can be produced on each
axis will depend on the rate of leaf formation *in vitro*; the rate at
which they can be induced to develop prematurely can to some extent be
controlled by the concentration of cytokinin, according to the sensitivity of the species. Serial subculture is generally carried out every
4-8 weeks. Multiplication rates vary widely but x5 to x10 per month is
not unusual.

Proliferating axillary shoot cultures may be started either from
cultured excised shoot tips or *in vitro* adventitious shoots. Shoot
tips (1-5 mm high) should be dissected from terminal or lateral buds.
The size need only be small enough to avoid contamination; the larger
the tip cultured the greater its chances of survival. For virus
elimination the smallest piece is used that will survive and develop.
This is referred to as a meristem-tip and usually comprises only the
apical dome and one or two leaf primordia (Walkey, this volume).

Shoot tips are placed on a medium with a low level of cytokinin
(eg. 0.05-0.5 mg/l BAP) and auxin (eg. 0.01-0.1 mg/l IBA). The level
of cytokinin should then be progressively raised at each subculture
until a satisfactory rate of branching can be achieved without excessive leaf distortion and yellowing. Auxin should be omitted from the
medium if possible, especially if there is any tendency for callus to
form. The presence of cytokinin with low or no auxin generally
inhibits root formation, a useful side effect which enables branches
to be separated more easily at each serial subculture. Rooting usually
occurs on transfer to a medium lacking cytokinin.

Enhanced axillary shoot production can now be carried out with many
different plants and is theoretically applicable to any species that
produces regular axillary meristems and responds to an available
cytokinin. Representative examples are *Gerbera* spp. (Murashige *et al.*,
1974), strawberry (Boxus, 1974), apple (Jones *et al.*, 1977), sugar beet
(Hussey & Hepher, 1978) and *Gladiolus* spp. (Hussey, 1977).

Adventitious shoots direct on explants

In some plants, particularly members of certain groups of monocotyledons, very few or occasionally no axillary meristems are formed.
Where this is so propagation depends mainly or entirely on the
formation of adventitious shoots from organ explants, either direct
or via callus. Natural regenerative behaviour or conventional
horticultural practice may suggest the most appropriate organ for
explants (eg. leaf pieces in *Begonia* spp.); failing this, all the main
organs should be tried, preferably the youngest available.

In most species adventitious shoots are induced by a high ratio of cytokinin to auxin in the culture medium, but there are considerable variations and some exceptions. In *Hyacinthus* spp. shoot formation on leaf and stem explants is triggered by low concentrations of auxin (0.1 mg/l NAA) and the addition of cytokinin has little further effect (Hussey, 1975). In *Brassica napus* (Kartha *et al.*, 1974) and *Beta vulgaris* (Hussey & Hepher, 1978) shoots are induced by cytokinin alone (0.25-1.0 mg/l BAP).

It should be noted that the amount of auxin required to trigger cell division in the explant may also inhibit shoot initiation and promote callus formation. It may be more effective, therefore, to begin with an approximately equal ratio of cytokinin to auxin and transfer to higher ratios in one or more steps at intervals of a few days or weeks.

Organ explants may give rise to one or more clusters of adventitious shoots, the development of which may inhibit further shoot induction on the same explant. If the first crop of shoots is insufficient, more may be obtained by the following two methods.

1. Allow the shoots to develop *in vitro* to the stage at which they can be used as a source of secondary explants. Shoots may be dissected into the necessary components such as stem or leaf pieces or, as with some monocotyledons, split vertically to destroy the main meristem and remove apical dominance.
2. Regularly trim the developing shoot clusters to a height of only a few mm above the original explant tissue. This may encourage further adventitious shoots to develop on the basal tissue, which continues to expand laterally.

Adventitious shoots from callus or cell suspensions

Callus is an unorganized mass of proliferating cells that can be induced on explants of various organs from almost any plant. An auxin is generally required for callus induction, often in combination with cytokinin. The effective concentrations of hormones vary considerably according to the species, but tend towards the higher ranges. The stronger and more stable synthetic auxins such as NAA are usually required, and in some plants 2,4-D is necessary to induce fast growing callus.

Organogenesis occurs in callus when it is transferred to media with less auxin, with or without an increase in the cytokinin level. Regenerated plants may arise from adventitious embryos, shoot apices or intermediate types of meristematic structure. Many calluses, especially those grown on higher hormone concentrations, become friable and can be broken up into suspensions of proliferating clumps of cells and single cells on transfer to agitated liquid medium (Street, 1977). These cell suspensions may produce adventitious embryos direct or give rise to regenerative callus after transfer to solid media.

With certain exceptions, the use of current techniques of callus and cell suspension culture for propagation must be considered only as a last resort, because most calluses have a propensity to accumulate genetically aberrant cells and not all species readily produce toti-potent callus. The real advantage of callus and cell suspensions lies in the comparative ease with which large quantities of tissue can be bulked up prior to plantlet formation. With proliferating shoot systems greater limitations are imposed on mass propagation by the labour intensive manipulations required to separate and subculture successive generations of shoot clusters. Techniques of mass embryo-genesis, so far established in only a few species, would seem to hold considerable promise for the future when some formidable technical problems have been overcome.

Mixed cultures

In practice the distinction between the three methods of shoot production outlined in the previous sections may not be clear cut and combinations of two or all three phenomena may occur to varying extents; should this happen, even if only for a short phase, the chances of achieving genetic uniformity may be reduced. Cultures should therefore be scrutinized regularly for the following possibilities.
1. Adventitious shoots and/or callus may form at the base of an otherwise normal looking axillary shoot cluster. This is usually the result of too high a concentration of cytokinin or auxin. Where possible auxin should be omitted from the medium as it may not be necessary for well established cultures. Hard, slow growing callus clearly not giving rise to adventitious shoots may be left.
2. If the process of enhanced axillary shoot formation is pushed too far by supra-optimal cytokinin concentrations thick clusters of shoots may be formed which include adventitious structures arising from distorted leaf or stem tissue and which may ultimately fuse into a callus-like mass.
3. Adventitious shoots may be accompanied by callus formation. This again may be an indication that the hormone levels are too high, but in many species some callus formation is inevitable due to the sensitivity of the explant tissues to conditions of culture. Callus arising at cut ends of explants may parallel the induction of shoots elsewhere and can be rejected on subculture. Where callus appears to envelop adventitious shoots it could easily be an extension of the initial cell division activity leading to direct regeneration of meristems within the explant tissue.

Storage of shoots

By lowering the temperature to 1-10°C according to the species, the growth of proliferating shoots may be slowed down so that the period between subcultures is extended to months or years. Stored cultures should contain a generous amount of medium and be sealed with thin polyethylene film to prevent excessive drying out. Very low light intensities will suffice (250 lx). Modifications to the medium, including an increase in the sucrose concentration from 3 to 10% and the addition of growth inhibitors or retardants, may improve the

survival rate when the shoots are recultured at normal temperatures
(Henshaw *et al.*, this volume).

Recalcitrant species

Numerous species have now been cultured successfully, but there remain
many in which either no response or only a poor reaction can be
obtained. Although there is no obvious reason why media cannot be
devised to suit the metabolic characteristics of each species, the
empirical nature of the approach often makes this a slow and pains-
taking process. Standard salt mixtures may require modification with
regard to one or more of the macro- or micro-nutrients and the
beneficial effect of extracts of the same or unrelated species may
indicate the need for metabolites or growth active substances not
usually included in basic media.

ROOTING AND PLANTING OUT

Although adventitious embryos already have a root initial and develop
direct into seedling-like plantlets, *in vitro* shoots multiplied on
medium containing cytokinin are generally without roots. Rooting
should be effected by transferring shoots to medium with low or no
cytokinin, with or without a rooting hormone such as IBA at concentrations of 0.1-1.0 mg/l or more. An alternative technique is to plant
in vitro shoots into sterile 15-20 mm cubes of plastic foam soaked in
rooting medium. When roots begin to protrude from the base of the foam
the cubes should be washed free of medium and planted into compost.

Tissue culture plantlets are equivalent to small rooted cuttings,
and transfer to conventional growing medium involves adaptation of both
shoot and root systems to normal conditions. The optimum stage for
transplanting is when the roots are beginning to grow out and the
developing leaves are becoming photosynthetically self-supporting.
An increase in the light intensity some days or weeks before transplanting increases the chances of survival; during this time provision
should be made for adequate entry of CO_2 by slackening screw-top or
other covers on the containers. After planting, an increase in the
humidity from mist or by polyethylene cover is advisable until the
roots have adapted to soil.

Plants that form resting organs may be allowed to produce these
in vitro under appropriate conditions as an alternative to direct
planting. The resulting tubers, cormels, bulbils or rhizomes make
convenient and easily transportable planting material. Cold treatment,
washing or hormone application may be necessary in some temperate
species to break dormancy.

REFERENCES

Boxus P. (1974)   The production of strawberry plants by *in vitro* micro-
propagation. *Journal of Horticultural Science* 49, 209-10.
Debergh P. (1977)   Symposium on tissue culture for horticultural
purposes. *Acta Horticulturae* 78.
Gamborg O.L., Murashige T., Thorpe T.A. & Vasil I.K. (1976)   Plant
tissue culture media. *In vitro* 12, 473-8.
Hussey G. (1975)   Propagation of hyacinths by tissue culture. *Scientia
Horticulturae* 3, 21-8.
Hussey G. (1977)   *In vitro* propagation of Gladiolus by precocious
axillary shoot formation. *Scientia Horticulturae* 6, 287-96.
Hussey G. (1978)   The application of tissue culture to the vegetative
propagation of plants. *Science Progress, Oxford* 65, 185-208.
Hussey G. & Hepher A. (1978)   Clonal propagation of sugar beet plants
and the formation of polyploids by tissue culture. *Annals of Botany*
42, 477-9.
Jones O.P., Hopgood M.E. & O'Farrell D. (1977)   Propagation *in vitro*
of M.26 apple rootstocks. *Journal of Horticultural Science* 52,
235-8.
Kartha K.K., Gamborg O.L. & Constabel F. (1974)   *In vitro* plant
formation from stem explants of rape (*Brassica napus* cv. *Zephyr*).
Physiologia Plantarum 31, 217-20.
Murashige T. (1974)   Plant propagation through tissue cultures. *Annual
Review of Plant Physiology* 25, 135-66.
Murashige T., Serpa M. & Jones J.B. (1974)   Clonal multiplication of
Gerbera through tissue culture. *HortScience* 9, 175-80.
Murashige T. & Skoog F. (1962)   A revised medium for rapid growth and
bioassays with tobacco tissue cultures. *Physiologia Plantarum* 15,
473-97.
Reinert J. & Bajaj Y.P.S. (1977)   *Applied and Fundamental Aspects of
Plant Cell, Tissue and Organ Culture*. Springer-Verlag, New York.
Skirvin R.M. (1978)   Natural and induced variation in tissue culture.
Euphytica 27, 241-66.
Skoog F. & Miller C.O. (1957)   Chemical regulation of growth and organ
formation in plant tissues cultured *in vitro*. *Symposia of the
Society for Experimental Biology* 11, 118-31.
Street H.E. (1977)   *Plant Tissue and Cell Culture*, 2nd edition.
Blackwell Scientific Publications, Oxford.
Thorpe T.A. (1978)   *Frontiers of Plant Tissue Culture, 1978*. Inter-
national Association for Plant Tissue Culture, Calgary.

Cryopreservation of plant cell and tissue cultures

L.A. WITHERS

Friedrich Miescher-Institut, PO Box 273,
CH-4002 Basel, Switzerland, and
Botanical Laboratories, University of Leicester,
Leicester LE1 7RH

INTRODUCTION

In recent years there has been an upsurge of interest in the development
of storage methods for plant tissue cultures. Several factors have
contributed to this. Threats to world germplasm resources, in the
form of natural selection and the destruction of natural habitats,
compounded by the trend towards concentration upon a low number of high
yielding varieties in horticulture and agriculture, require that
positive action be taken to conserve the full spectrum of genetic
diversity now extant (Frankel & Hawkes, 1975). Although conventional
seed storage methods will satisfy this requirement for many species,
an alternative method is required for vegetatively propagated plants
and species with recalcitrant seed. Possible solutions to the problem
are *in vitro* propagation and storage of the former, and the development
of novel seed/embryo storage methods, or even introduction into culture
of explants, for the latter.

Plant breeders who wish to utilize tissue culture methods for
clonal propagation, variety improvement, pathogen eradication or the
development of new genomes through somatic cell genetics (Murashige,
1978; Walkey, 1978; Thomas *et al.*, 1979) clearly require a method for
the maintenance of valuable stock tissues. Similarly, those who use
tissue cultures for physiological and biochemical studies, or for
experimental genetic manipulation, also require reliable methods for
storage of their cell lines. Expense, the possibility of loss through
human error, equipment failure and the ever-present risk of genetic
change with time in culture (D'Amato, 1978), militate against the use
of serial subculturing for long-term culture maintenance.

Many of these problems could be reduced by controlling the normally
rapid rate of growth in culture. Suppression of growth would lead to
lower labour and materials expenditure whilst at the same time reducing
the accumulation of genetic abnormalities. Several technical approaches
have been attempted, including desiccation of the tissues, growth
under reduced atmospheric pressure or partial pressure of oxygen and,
most successfully, growth at reduced temperatures (Withers, 1980).

Further consideration of aspects of minimal growth storage is given by Henshaw *et al*. (this volume). The complete suspension of growth is, however, the most desirable maintenance condition, and this can only be achieved by storage of tissues at ultralow temperatures.

Successful cryopreservation of living material was first reported some 30 years ago but it was not until 1968 that the technique was extended to include cultured plant tissues (Withers & Street, 1977a). Since that time rapid progress has been made and it is now possible to preserve either cell cultures, callus, haploid pollen embryos, protoplasts, somatic embryos and plantlets, or shoot tip meristems, of about 30 plant species, as well as seed and pollen of many more (Frankel & Hawkes, 1975; Bajaj & Reinert, 1977; Withers & Street, 1977a; Withers, 1980). Nonetheless, those familiar with the vast genotypic scope encompassed by plant tissue culture will realize that this success represents only a beginning. The earliest studies concentrated upon model systems such as suspension cultures of *Daucus carota* and *Acer pseudoplatanus,* and it is only recently that cryopreservation methods have been applied to specimens of importance in genetic conservation, plant breeding and somatic cell genetics (e.g. Grout & Henshaw, 1978; Withers & King, 1979a).

There are two possible broad approaches to cryopreservation, namely by rapid freezing or by slow freezing. In the former, the specimen is protected against freezing damage by rapid transfer to the storage temperature. Ice crystals which form in the cells are of a microscopic size and do not disrupt cellular ultrastructure provided that thawing is carried out rapidly enough to prevent recrystallization. However, for many specimens, including suspension cultures, rapid freezing has not proved successful and the alternative approach must be used. This depends upon extracellular freezing for protection. As the cell is cooled slowly, ice forms outside it and the consequent drop in water vapour pressure causes a flow of water from the protoplast. This process of protective dehydration reduces the amount of freezable water in the cell, thereby preventing ice damage upon subsequent transfer to ultralow temperatures. Effective dehydration can also be induced by prefreezing or stepwise freezing (Withers, 1978a), in which the specimen is held at a specific sub-zero temperature before transfer to storage. Since there are deleterious effects of dehydration and intracellular solute concentration (solution effects) over-dehydration should be avoided.

Both rapidly and slowly frozen specimens normally require the application of chemical cryoprotectants prior to freezing. These compounds protect against both ice and solution effect damage. For further discussion of cryopreservation in general, freeze-thaw phenomena, freezing damage and cryoprotectant action, see Meryman (1966), Mazur *et al*. (1972) and Withers (1980).

This chapter deals mainly with the cryopreservation of suspension cultures and somatic embryos and plantlets derived from them. Despite the successes reported here it is clear that the factors limiting the wider application of cryopreservation methodology are still not fully understood. It is important that attention be devoted in the future to further elucidating the roles played by the culture conditions prior to freezing and after thawing, to optimizing freeze-thaw protocols for many important specimen types, and to understanding the nature of cryodamage in cultured plant material. It should also be noted that cryopreservation of many specimens must await the development of adequate tissue culture systems, since a specimen which grows poorly *in vitro* is unlikely to respond favourably to the stresses imposed upon it by ultra-low temperature storage.

MATERIALS AND METHODS

This section contains freeze-thaw protocols, treatments which help to render the specimen more suitable for preservation, treatments which aid recovery and, finally, methods for assessing the condition of the preserved specimen.

Pregrowth

The stage of the culture growth cycle is critical. Suspension cultures should be in the late lag or exponential phase of growth; stationary phase cells are entirely unsuitable (Withers & Street, 1977b). Somatic embryos should be in the early stages of development. Small structures will usually respond more favourably to freezing and thawing than larger ones. Small, highly cytoplasmic cells typical of a rapidly growing culture are most resistant to cryodamage. In species which normally have large cells, e.g. *Rosa* sp. and *A. pseudoplatanus,* the mean cell size can be reduced by rapid subculturing (e.g. every 7 days) and/or by growth in medium supplemented with mannitol at 3-6% w/v or proline at 5-10% w/v (Withers & Street, 1977b; Withers & King, 1979a). Pregrowth at reduced temperatures is not normally beneficial for suspension cultures, although an extended period of exposure to cryoprotectants at normal or reduced temperatures may be of some benefit (Withers & Street, 1977b; Kartha *et al.,* 1979; Withers, 1980).

Cryoprotection and preparation for freezing

A wide range of compounds have cryoprotectant properties. Dimethyl-sulphoxide (DMSO) and glycerol are most commonly used alone, together, or in combination with other compounds such as sugars, sugar alcohols and amino acids. Suitable cryoprotectants for cell suspensions are solutions with final concentrations of: 5-10% w/v DMSO; 10% w/v proline; 5% DMSO and 10% w/v glycerol; 5% DMSO, 5% glycerol and 10% w/v proline or 33% w/v sucrose; the latter is highly viscous and difficult to filter, but is very effective with a range of suspension cultures (Withers & King, unpublished). DMSO alone may be most suitable for protoplasts, meristems, zygotic and somatic embryos and callus pieces. The chosen cryoprotectant is prepared at twice the final concentration in culture medium, sterilized, preferably by filtration, and then chilled on ice. It is added to an equal volume of the chilled specimen in several aliquots over a period of *c*. 1 h (or in one addition if tolerance can be demonstrated), and the cryoprotected specimens are left on ice for a further hour before freezing. The cryoprotected cells are dispensed into sterile containers, e.g. 1 ml into a 2 ml screw-top polypropylene ampoule, and the ampoules carefully labelled (by engraving with a sharp needle if an alcohol bath is to be used for freezing). Other specimens such as callus pieces, embryos and plantlets may be treated as suspensions. Alternatively, prepare them for dry-freezing (Withers, 1978b) by filtering them or lifting them from the medium with sterile forceps, blotting them dry on sterile filter paper and enclosing them in an aluminium foil envelope. Pollen and dry seed may be enclosed in foil without any cryoprotection.

Freezing

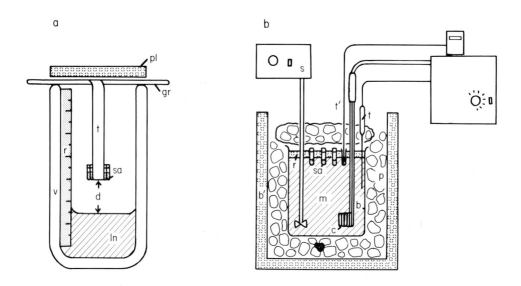

Figure 1. Improvised freezing apparatus. a, The specimen ampoules (sa) are suspended from a glass rod (gr) by tape (t) at a known distance (d) above the surface of liquid nitrogen (ln) in a vacuum flask (v). A rule (r) is used to locate the specimen and a loose fitting polystyrene lid (pl) is placed over the neck of the flask. b, The specimen ampoules (sa) are inserted through holes in a polystyrene raft (r) floating on methanol (m) in a large beaker (b). Polystyrene chips (p) are packed between the beaker and an enclosing plastic bin (b'). A bag of polystyrene chips is placed over the beaker and its contents. The methanol is cooled by a thermostati-cally controlled, electrically powered cooling coil (c). A resistance thermometer (t) transmits the methanol bath temperature to the cooling unit and a second thermometer (t') records the temperature of a dummy specimen ampoule. A stirrer (s) ensures circulation of the methanol.

Either slow or stepwise freezing can be used. In the first case freezing at 1° or 2°C/min to -100°C is a suitable starting point. If unsuccessful, slower rates down to 0.1°C/min or more rapid rates up to about 10°C/min can be tried. Freezing slowly beyond an intermediate sub-zero temperature before transfer to ultralow temperatures is unnecessary and sometimes deleterious. Although commercial linear cooling units are available (Withers, 1980), satisfactory and repro-ducible non-linear cooling rates can be achieved using the simple apparatus illustrated in Figure 1a. A similar result can also be achieved by placing the specimen ampoules inside an insulated container within a deep freeze (-70° or -100°C; Sala *et al.*, 1979).

Stepwise freezing involves the use of a coolant bath such as methanol cooled with solid CO_2 or an electrically powered cooling coil. The specimen may be held in a bath at -30° or -40°C for *c*. 40 min, or the more elaborate apparatus illustrated in Figure 1b may be used. The latter provides cooling to a chosen holding temperature at a reproduci-ble rate (e.g. -35°C in 40 min) and maintenance of that temperature for the required length of time (e.g. 30 min). Such a freezing programme has been used by Withers & King (1979a, 1979b, and unpublished) to preserve a range of suspension cultures.
Note: Alcohol bath freezers cannot be used for dry-frozen specimens unless they are enclosed in an impermeable vial.

Storage

Once the specimen has reached the desired temperature, it should be rapidly transferred to liquid nitrogen in a Dewar vessel. This will suffice for very short term storage, but a purpose built liquid nitrogen refrigerator (e.g. Union Carbide LR-40) is best for secure, organized, long term storage. Electrical deep freezers running at *c*. -20°C are entirely unsuitable, although a storage temperature of -70° to -100°C may be adequate for a number of days or weeks.

Thawing

Slowly frozen specimens, in particular suspension culture cells or somatic embryos which have been dehydrated sufficiently during freezing, may occasionally be thawed slowly without loss of viability by being stood in air at room temperature (Withers, 1978b). However, the majority of slowly frozen specimens must be thawed rapidly, since some freezable water usually remains in the cells. Specimens frozen in ampoules should be agitated in sterile warm water (*c*. +40°C) until the majority of ice in the suspending medium has melted. Dry-frozen specimens, however, may suffer if exposed to liquids at this stage and should be thawed by rapid transfer with forceps to semi-solid medium (Withers, 1978b).

Post-thaw procedures

Suspension cultures in liquid medium may simply be dispensed into a

larger volume of fresh medium (e.g. 1 ml into 5 ml) and placed on a
platform shaker. This treatment can result in unnecessary loss of
viability, as may post-thaw washing before return to culture (Withers &
King, 1979a; cf. Withers & Street, 1977b; Sala *et al.*, 1979). If
damage is suspected, the specimen should be layered over semi-solid
medium in a Petri dish and left in static culture until there is
adequate growth of new tissue for initiation of a suspension culture
or for transfer to further semi-solid medium. The choice of post-thaw
culture medium constituents will depend upon the desired pattern of
growth. Thus, for continued unorganized growth a normal maintenance
medium should be used. For the induction of organized growth (or, in
the case of somatic embryos and plantlets, and meristems, for the con-
tinuation of organized growth) the appropriate promotory medium should
be used; the addition of activated charcoal (Withers, 1978b) or
hormones such as gibberellic acid (Grout *et al.*, 1978) may be beneficial.

Assessment of survival and recovery growth

The condition of freshly thawed specimens can be ascertained by using
viability tests such as the following (see Withers, 1980, for a
critical discussion).

Fluorescein diacetate (FDA) staining   Add 0.1 ml of a 0.5% w/v stock
solution of FDA in acetone to 5 ml of fresh culture medium. Mix one
drop of this with one drop of a cell suspension culture and observe in
a light microscope under ultraviolet illumination. Viable cells will
fluoresce brightly.

Evan's Blue staining (complementary to FDA staining)   Add one drop
of a 0.025% w/v solution of Evan's Blue in culture medium to one drop
of cell suspension and observe in a light microscope. Viable cells
will exclude the stain.

Tetrazolium chloride reduction   Add 1% w/v solution of 2,3,5-tri-
phenyl tetrazolium chloride (TTC) in 0.05 M phosphate buffer (pH 7.5)
to an equal volume of cell suspension (or to a bulk specimen in liquid
medium) and incubate the mixture at 20°C for *c.* 15 h in the dark.
Viable cells reduce the TTC to formazan and stain red. To estimate
viability quantitatively extract the formazan as follows: wash the
cells twice with distilled water and then incubate at *c.* 75°C in 95%
ethanol. The absorption of the extract at 485 nm can be compared with
the standard curve constructed using artificial mixtures of live and
dead cells.

   During recovery growth, the percentage viability can be determined
at intervals. Also, biomass estimations (packed cell volume, cell
density, dry weight) and mitotic index measurements can be made (Street,
1977). In order to observe the subcellular effects of freezing and
thawing, and to follow the progress of recovery growth at the ultra-
structural level, electron microscopical observation can be carried out
using standard fixation techniques and low temperature techniques such
as freeze-substitution and freeze-etching (Withers, 1980).

CONCLUSIONS

In conjunction with the methods described by Henshaw *et al.* (this volume) the above are offered as guidelines for the development of suitable freeze-preservation protocols for tissue culture specimens. Unfortunately the approach must, as yet, be empirical. However, progress in recent years in adapting existing methods to a wider range of specimens presages well for the future. It may soon become feasible to construct a methodological key from which the most suitable combination of experimental parameters may be determined according to the known characteristics of the specimen.

REFERENCES

Bajaj Y.P.S. & Reinert J. (1977) Cryobiology of plant cell cultures and establishment of gene banks. *Applied and Fundamental Aspects of Plant Cell, Tissue and Organ Culture* (Ed. by J. Reinert & Y.P.S. Bajaj), pp. 757-77. Springer-Verlag, Berlin.
D'Amato F. (1978) Chromosome variation in cultured cells and regenerated plants. *Frontiers of Plant Tissue Culture, 1978* (Ed. by T.A. Thorpe), pp. 287-96. International Association for Plant Tissue Culture, Calgary.
Frankel O.H. & Hawkes J.G. (1975) *Crop Genetic Resources for Today and Tomorrow.* University Press, Cambridge.
Grout B.W.W. & Henshaw G.G. (1978) Freeze-preservation of potato shoot-tip cultures. *Annals of Botany* 42, 1227-9.
Grout B.W.W., Westcott R.J. & Henshaw G.G. (1978) Survival of shoot meristems of tomato seedlings frozen in liquid nitrogen. *Cryobiology* 15, 578-83.
Kartha K.K., Leung N.L. & Gamborg O.L. (1979) Freeze-preservation of pea meristems in liquid nitrogen and subsequent plant regeneration. *Plant Science Letters* 15, 7-15.
Mazur P., Leibo S.P. & Chu E.H.Y. (1972) A two-factor hypothesis of freezing injury - Evidence from Chinese hamster tissue culture cells. *Experimental Cell Research* 71, 345-55.
Meryman H.T. (1966) *Cryobiology.* Academic Press, London, New York.
Murashige T. (1978) The impact of plant tissue culture on agriculture. *Frontiers of Plant Tissue Culture, 1978* (Ed. by T.A. Thorpe), pp. 15-26. International Association for Plant Tissue Culture, Calgary.
Sala F., Cella R. & Rollo F. (1979) Freeze-preservation of rice cells. *Physiologia Plantarum* 45, 170-6.
Street H.E. (1977) Cell (suspension) cultures-techniques. *Plant Tissue and Cell Culture* (Ed. by H.E. Street), pp. 61-102. Blackwell Scientific Publications, Oxford.
Thomas E., King P.J. & Potrykus I. (1979) Improvement of crop plants via single cells *in vitro* - an assessment. *Zeitschrift für Pflanzenzüchtung* 82, 1-30.
Walkey D.G.A. (1978) *In vitro* methods for virus elimination. *Frontiers of Plant Tissue Culture, 1978* (Ed. by T.A. Thorpe), pp. 245-54. International Association for Plant Tissue Culture, Calgary.

Withers L.A. (1978a) Freeze-preservation of cultured cells and tissues. *Frontiers of Plant Tissue Culture, 1978* (Ed. by T.A. Thorpe), pp. 297-306. International Association for Plant Tissue Culture, Calgary.

Withers L.A. (1978b) Freeze-preservation of somatic embryos and clonal plantlets of carrot (*Daucus carota* L.). *Plant Physiology* 63, 460-7.

Withers L.A. (1980) Low temperature storage of plant tissue cultures. *Advances in Biochemical Engineering,* 11 (Ed. by A. Fiechter). Springer-Verlag, Berlin (in press).

Withers L.A. & King P.J. (1979a) Proline – a novel cryoprotectant for the freeze-preservation of cultured cells of *Zea mays* L. *Plant Physiology* 64, 675-8.

Withers L.A. & King P.J. (1979b) The freeze-preservation of cultured plant cells. *Experientia* 35, 984.

Withers L.A. & Street H.E. (1977a) The freeze-preservation of plant cell cultures. *Plant Tissue Culture and its Bio-technological Application* (Ed. by W. Barz, E. Reinhard & M.-H. Zenk), pp. 226-44. Springer-Verlag, Berlin.

Withers L.A. & Street H.E. (1977b) The freeze-preservation of cultured plant cells. III. The pregrowth phase. *Physiologia Plantarum* 39, 171-8.

Tissue culture methods for the storage and utilization of potato germplasm

G.G. HENSHAW, J.F. O'HARA & R.J. WESTCOTT

Department of Plant Biology, University of Birmingham,
PO Box 363, Birmingham B15 2TT

INTRODUCTION

At least a proportion of the germplasm of vegetatively propagated crops
such as potato is stored in clonal form because of problems of low
fertility and heterozygosity. The lack of vegetative propagules
suitable for long-term storage in these crops means that it is necessary
to have an annual cycle of growth, which is expensive and increases the
chance of losses occurring as a result of diseases or other natural
causes. Alternative methods of germplasm storage based on tissue
culture methods are, however, now available. These methods have the
advantage that the same basic procedures can be employed both to
eliminate disease and to exclude pathogens from disease-free material;
further, high multiplication rates can be achieved when required.

Although most pathogens are excluded by the aseptic procedures which
are an essential feature of *in vitro* methods, there are some which are
not necessarily eliminated or even detected. In particular, viruses can
readily multiply in tissue culture. These can be eliminated by meristem
or shoot-tip culture, possibly in combination with heat therapy (Henshaw,
1979), although total success cannot be guaranteed. Efficient indexing
procedures are therefore essential. Once disease-free, or at least
pathogen-tested, cultures have been produced standard aseptic procedures
will prevent the reinfection of stored material. Such cultures can also
facilitate the utilization of germplasm by providing a system for the
international distribution of material for breeding programmes (Roca *et
al.*, 1979).

The methods adopted for long-term storage of germplasm must maintain
a satisfactory level of genetic stability as well as guaranteeing
survival. Although unorganized types of cultures (callus, suspension
cultures, etc.) readily survive, there are with many species problems of
genetic instability and loss of morphogenetic potential (D'Amato, 1975).
Shoot-tip cultures, employed in the same manner as for disease elimination and exclusion, provide a valid alternative means of storage for
many crop species; multiplication is based on the non-adventitious
production of axillary buds. The rates of multiplication which can be

achieved with shoot-tip cultures may be lower than those theoretically attainable with callus or suspension cultures, but they are usually more than adequate. Furthermore, experience with a number of horticultural species indicates that meristems produced *in vitro* retain the genetic stability which is characteristic of *in vivo* meristems.

An *in vitro* system with high multiplication rates is not ideal for germplasm storage because it requires frequent attention and the mutation rate is likely to correlate strongly with the rate of cell division. For this reason, in an ideal system cell division would be completely suppressed, but with present techniques this is only likely to be achieved on a long-term basis by storage at ultra-low temperatures, most conveniently obtained with liquid nitrogen (-196°C). Such a technique has been used successfully with plant cell cultures and, more recently, with isolated meristems (Seibert & Wetherbee, 1977; Grout & Henshaw, 1978), but it is not yet known whether it will prove to be generally applicable. Techniques based on the storage of shoot-tip cultures under conditions which permit only minimal rates of growth are likely to be more widely employed. These methods have the advantages that the stored material is readily available for use and that the stocks are replenished as the cultures grow. We present here detailed instructions for the techniques of both minimal growth storage and cryopreservation, using potato shoot-tip cultures as the starting material.

METHODS AND MATERIALS

Shoot-tip cultures

1. Surface sterilize vegetative shoots of potato by immersion in a 5% solution of sodium hypochlorite (0.6% available chlorine) containing a drop of Tween 80 detergent, for 10 min; rinse thoroughly in sterile distilled water.
2. Initiate cultures by aseptically excising the apical dome plus one to four leaf primordia from axillary buds. Place onto a strip of filter paper (1 x 7.5 cm) folded to form a bridge in each test tube containing 4 ml of liquid Murashige & Skoog (MS; 1962) medium supplemented with growth hormones (0.5 mg/l 1-naphthaleneacetic acid (NAA); 0.2 mg/l gibberellic acid (GA); and with or without 3.0 mg/l benzyladenine (BA) depending on the potato variety; Westcott *et al.*, 1977).
3. Incubate the cultures with a 16 h photoperiod of 4000 lx, at 22°C.
4. Single or multiple shoot cultures are produced which may be maintained in static or agitated liquid medium. They may be further propagated by taking nodal cuttings, placing on MS medium without growth hormones and incubating as in 3. The cultures should be subcultured every 4-12 weeks. To extend this passage length to 1 year, minimal growth storage techniques should be used, as described below.

Minimal growth storage

Starting with nodal or multiple shoot cultures minimal growth storage
has been successfully applied to a range of potato genotypes, including
diploid, triploid, tetraploid and pentaploid varieties (Table 1);
survival data are based upon regrowth of at least part of the culture on
transfer to fresh medium at the end of 12 months. Since genotypes vary
in their ability to withstand minimal growth conditions (Table 1) the
storage conditions must be tailored to suit the plant material. However,
the following is a good system with which to start.

Table 1. Percentage survival of single-node cultures from several
tuberous *Solanum* species stored for increasing periods in 3.5 ml of
basal MS medium + 3% sucrose at a continuous temperature (6°C).

Storage period (months)	Mean %	Species code* and Ploidy						
		GON 2X	TBR 4X	CHA 3X	STN 2X	ADG 4X	JUZ 3X	CUR 5X
6	73	31	61	93	72	81	83	91
9	58	21	36	61	64	65	75	85
12	49	14	20	56	59	51	63	77
18	29	0	0	0	39	31	61	71

*Key to code: GON, *Solanum goniocalyx*; TBR, *S. tuberosum* ssp.
tuberosum; CHA, *S. x chaucha*; STN, *S. stenotomum*; ADG, *S. tuberosum*
ssp. *tuberosum*; JUZ, *S. x juzepczuki*; CUR, *S. x curtilobum.*

Basal growth conditions consist of liquid MS medium containing 3%
sucrose and no hormones, the plant material being supported on a 3 x 10
cm filter paper bridge over 50 ml of medium in a screw-cap jar (12 cm
height), and incubated at 22°C with a 16 h photoperiod of 4000 lx.
Reduced growth rates can be achieved by adopting some or all of the
following modifications.

Physical conditions
   1. Reduce the temperature to 6-12°C. (If using the technique with
      tropical species, it should be noted that such low temperatures may
      be lethal, and use of the range 14-18°C should be adequate.)   In
      our laboratory, material routinely maintained under a regime of 6°C
      during an 8 h night and 12°C during a 16 h day has a mean survival
      of *c*. 80%.
   2. Keeping the temperature at 22°C, subject the cultures to a higher
      osmolarity by adding 3 or 6% mannitol to the medium.

Medium   Using low temperature conditions of 10°C, raise the sucrose concentration to 8%.

Addition of growth suppressants   Add abscisic acid (ABA) at 5.0-10.0 mg/l. Of the several compounds we have tested, ABA is used routinely in our laboratory.

Under suitable minimal growth conditions, survival rates are improved at 10°C by increasing the nutrients available by raising the volume of medium to 60 ml per jar. In addition, the use of polypropylene film allows a useful compromise between gas exchange and moisture retention, since the evaporation of medium during long storage periods can cause losses, yet sealing the vessel drastically reduces survival, due to prevention of gas exchange.

Shoot-tip storage in liquid nitrogen (see Fig. 1)

1. Surface sterilize as for shoot-tip cultures.
2. Excise the apical dome and one to two leaf primordia from axillary buds.
3. Place several such shoot-tips on filter paper floated on liquid MS medium supplemented with 1 mg/l BA. Incubate for 2 days at 22°C.
4. Prior to freezing, flood the Petri dish with 10 ml of a 10% solution of dimethyl sulphoxide (DMSO) cryoprotectant dissolved in MS medium plus BA. After 1 h arrange the shoot-tips separately on dissecting needles, then plunge them directly into a 6 cm depth of liquid nitrogen. They freeze immediately, sticking to the needle. They may be stored in liquid nitrogen at this stage for as long as desired.
5. Thaw frozen shoot-tips rapidly by plunging into medium at 34°C; they float free, the medium diluting out the cryoprotectant.
6. Place the thawed shoot-tips individually onto filter paper bridges in tubes of liquid medium. Incubate in a controlled environment of 22°C with continuous light of 4000 lx. Survivors regrow into plantlets. A typical experiment should include adequate controls to ensure that it is the response to ultra-fast freezing, rather than to cryoprotection or thawing, which is elicited.

Factors affecting storage   Within this basic procedure the following factors can be varied to optimize survival rates for particular species.
1. Age and physiological state of the whole plant and individual buds: material of various ages should be tried.
2. Shoot-tip size (hence the number of primordia): the volume of explant affects the cooling rate of the innermost cells.
3. Composition of the post-excision and post-thaw media: different genotypes have different requirements; also thawed explants are at least partly damaged, and this may alter their requirements.
4. Duration and environmental conditions of the post-excision period: this should be just sufficient to permit recovery from the trauma of excision without allowing too much growth of the explant.
5. Choice of cryoprotectant: cells differ in their water content, cytoplasmic density and surface area, and thus require different

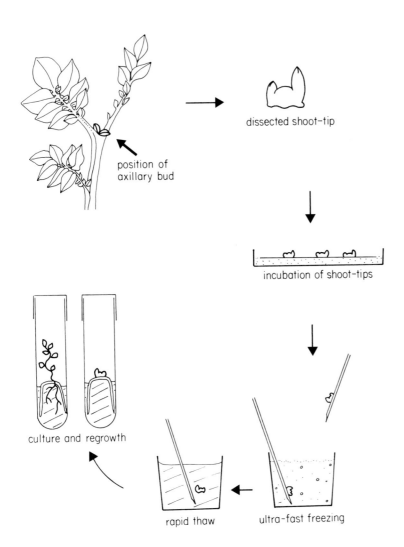

position of
axillary bud

dissected shoot-tip

incubation of shoot-tips

culture and regrowth

rapid thaw   ultra-fast freezing

Figure 1.   Procedure for cryopreservation of potato shoot-tips.

cryoprotectant treatments. DMSO penetrates rapidly (protects from
within the cell) and is non-toxic below 15% v/v; other cryoprotectants may act at the cell surface or osmotically (see Withers, this
volume), but each has an optimum effective treatment time.

6. Freeze/thaw procedure: the procedure given has been successful with
   Solanum goniocalyx (Grout & Henshaw, 1978; Stamp, 1978), *S. tuber-*
   osum ssp. *andigena*, and a few cultivars of *S. tuberosum* ssp.
   tuberosum (Stamp, 1978, and unpublished data), with survival rates

of 54%, 36% and 5-40% respectively. Carnation meristems (Seibert & Wetherbee, 1977) have also been frozen in a similar way, but thawed more slowly. However other species tested have not survived ultra-fast freezing and it may only be suitable for a limited number of species having cells with certain (as yet undetermined) character-istics. Nevertheless, this procedure certainly merits a trial in any freezing programme because of its speed and simplicity.

REFERENCES

D'Amato F. (1975) The problem of genetic stability in plant tissue and cell cultures. *Crop Genetic Resources for Today and Tomorrow* (Ed. by O. Frankel & J.G. Hawkes), pp. 333-47. University Press, Cambridge.

Grout B.W.W. & Henshaw G.G. (1978) Freeze preservation of potato shoot-tip cultures. *Annals of Botany* 42, 1227-9.

Henshaw G.G. (1979) Plant tissue culture: its potential for dissemina-tion of pathogen-free germplasm and multiplication of planting material. *Plant Health* (Ed. by D.L. Ebbels & J.E. King), pp. 139-47. Blackwell Scientific Publications, Oxford.

Murashige T. & Skoog F. (1962) A revised medium for rapid growth and bioassays with tobacco tissue cultures. *Physiologia Plantarum* 15, 473-97.

Roca W.M., Bryan J.E. & Roca M.R. (1979) Tissue culture for the international transfer of potato genetic resources. *American Potato Journal* 56, 1-10.

Seibert M. & Wetherbee P.J. (1977) Increased survival and differentia-tion of frozen herbaceous plant organ cultures through cold treatment. *Plant Physiology* 59, 1043-6.

Stamp J.A. (1978) Freeze preservation of shoot-tips of potato varieties in liquid nitrogen. *M.Sc. Thesis, University of Birmingham.*

Westcott R.J., Grout B.W.W. & Henshaw G.G. (1977) Tissue culture storage of potato germplasm: culture initiation and plant regenera-tion. *Plant Science Letters* 9, 309-15.

Section 3
Methods for use with viruses

The infection of cucumber protoplasts with cucumber mosaic virus (CMV) or viral RNA

K.R. WOOD, M.I. BOULTON & A.J. MAULE

Department of Microbiology, University of Birmingham,
PO Box 363, Birmingham B15 2TT

INTRODUCTION

Plant protoplasts offer a valuable approach to the study of many aspects of the virus/host interaction. Since the original observations by Cocking (1966) on the infection of tomato fruit protoplasts by tobacco mosaic virus and the subsequent critical extension of the technique by Takebe (Takebe & Otsuki, 1969) to the infection of tobacco leaf mesophyll protoplasts, the procedure has been applied, with modifications, to many combinations of host and virus. Refinements in techniques for preparing and maintaining protoplasts, and for their efficient infection, have led to the elucidation of events in the replication cycle of several viruses which would be difficult, if not impossible, to study in the intact plant. Examples have been comprehensively reviewed (Zaitlin & Beachy, 1974; Takebe, 1975, 1977).

Although attention has inevitably focused on the virus replication cycle, the potential for the use of protoplasts in a wider context is increasingly being realized. They have already been used, for example, in observations on the phenomena of cross protection (Otsuki & Takebe, 1976) and hypersensitivity (Otsuki et al., 1972). Our interest has centred on the use of protoplasts in the study of resistance to virus infection (Wood et al., 1979), and to this end we have developed procedures for the preparation and maintenance of protoplasts from cultivars of cucumber, either resistant or susceptible to cucumber mosaic virus (CMV) (Coutts & Wood, 1977a; Maule et al., 1979a) and for their infection with CMV in the presence of poly-L-ornithine (PLO) (Coutts & Wood, 1976, 1977b; Maule et al., 1979a). As a consequence, it is possible to study events in the infection process and in the virus replication cycle in both resistant and susceptible cells.

Interest in the early events of virus infection has also led us to achieve successful infection of cucumber protoplasts with CMV-RNA (Maule et al., 1979b). Modification of the procedure employing polyethylene glycol as an uptake inducer (Dawson et al., 1978), a technique which proved superior to others for this virus/host combination (Sarkar et al., 1974; Beier & Bruening, 1976; Mühlbach et al., 1977), permits routine

Ingram D.S. & Helgeson J.P. (1980) *Tissue Culture Methods for Plant Pathologists*

infection more efficiently and reproducibly than procedures using virus/
PLO complexes. Infection with tobacco ringspot virus can also be
achieved by the same method. Successful infection with viral RNA also
provides the opportunity, therefore, to study the biological role of the
individual RNA species of multicomponent viruses by introduction of the
genome segments either singly or in combination.

   Here we describe a one-step method for the preparation of protoplasts
from a susceptible cucumber cultivar (*Cucumis sativus* L. cv. Ashley) and
for their infection by CMV-W and its RNA.

ORGANISMS

Plants

Cucumis sativus L. cv. Ashley plants should be germinated and grown in
Arthur Bowers' potting compost (Lindsey Kesteven Fertilizers, Saxilby,
Lincoln) in 9 cm pots in a constant environment cabinet at 20°C with a
14 h photoperiod of 12 000 lx; use fully expanded first true leaves.
Vigna sinensis Endl. cv. Blackeye plants should be germinated and grown
in Arthur Bowers' potting compost in 9 cm pots in the glasshouse at
22°C ± 3°C with a 14 h photoperiod of 8500 lx; use when primary leaves
are fully expanded (*c*. 10 days).

Virus

The W strain of cucumber mosaic virus should be maintained in and
purified from *Nicotiana tabacum* L. cv. White Burley by the method of
Scott (1963) using two cycles of low and high speed centrifugation.

METHOD

*1. Isolation of protoplasts (Maule et al., 1979a, after Coutts & Wood,
1977a)*

Reagents required
Enzyme solution: 0.14% Driselase (Kyowa Hakko Kogyo Co. Ltd, Japan)
previously desalted on Sephadex G25; 0.5% pectinase (Sigma Chemical Co.,
London); 0.5% potassium dextran sulphate (Meito Sangyo Co. Ltd, Japan);
all in osmotically stabilized salts solution (SS) consisting of mannitol
(10%), sucrose (1%), KH_2PO_4 (0.1 mM), KNO_3 (0.5 mM), $CaCl_2$ (5 mM), $MgSO_4$
(0.5 mM), KI (0.5 µM), $CuSO_4$ (5 nM), adjusted to pH 5.8 with 0.1 M NaOH.
10% mannitol.
Exclusion dye for the assessment of viability: 0.1% phenosafranine in
10% mannitol + 0.1 mM $CaCl_2$.

Procedure   All operations should be carried out in a laminar flow
cabinet.

a. Surface sterilize fully expanded first leaves of cucumber plants
   by dipping into 70% ethanol, soaking for 10 min submerged in 4%
   Chloros + 0.03% Teepol, and then washing in three changes of
   sterile distilled water, allowing the leaves to stand in the third
   wash.
b. Peel the lower epidermis from sterile leaves and plasmolyse by
   placing the pieces peeled side down onto osmotically stabilized
   salts solution (SS) in Petri dishes. Incubate in the dark for 20
   min.
c. Pipette off the SS and filter sterilize 20 ml of the enzyme
   solution into each Petri dish containing c. 1.5 g of peeled tissue.
d. Incubate in the dark at 25°C on an orbital shaker (60 rev/min) for
   approximately 4 h or until the release of protoplasts with only
   slightly more vigorous manual shaking is achieved.
e. Filter the digested tissue through nylon mesh (68 µm; Henry Simon
   Ltd, Stockport, Cheshire) into 50 ml glass centrifuge tubes. Wash
   the protoplast suspension three times with mannitol, with sedimentation between washings at 100 g for 3.5 min.
f. Assess viability by gently mixing equal volumes of protoplast
   suspension and stain; examine and count using a modified Fuchs
   Rosenthal haemocytometer.

*2. Inoculation of cucumber protoplasts with viral nucleoprotein (Maule
et al., 1979a)*

Reagents required
40 µg Poly-L-ornithine (MW 122 000 or 270 000; Sigma Chemical Co.,
London) in 8 ml osmotically stabilized 50 mM potassium phosphate buffer,
pH 5.7.
40 µg CMV-W in 2 ml osmotically stabilized 50 mM potassium phosphate
buffer, pH 5.7, held at 0°C.
10% mannitol.
10% mannitol in 0.1 mM $CaCl_2$.
Osmotically stabilized salts solution (SS; see *1*).
Carbenicillin (0.4 g/ml; Pyopen, Beecham Research Laboratories,
Brentford).
Mycostatin (5×10^4 units/ml; E.R. Squibb & Sons Ltd, London).

Procedure
a. Incubate the virus suspension and poly-L-ornithine together in a 50
   ml glass centrifuge tube at 25°C for 5 min. Place the tube into
   ice and, when cool, pour onto the pelleted protoplasts (1 ml; 1.0-
   3.0×10^7). Add 9 ml mannitol and incubate in ice for 10 min, then
   centrifuge at 100 g for 3.5 min.
b. Repeat step a. twice more.
c. Wash the inoculated protoplasts three times in mannitol + $CaCl_2$ and
   resuspend in 20 ml SS containing 20 µl of each of the two antibiotics.
d. Incubate the protoplasts in 10 ml volumes in 100 ml conical flasks
   at 22°C with continuous diffuse lighting for approximately 38 h.

*3. Inoculation of cucumber protoplasts with viral RNA (Maule et al.,
1979b after Dawson et al., 1978)*

Reagents required
0.2 ml CMV-RNA (0.5 mg/ml) prepared from purified virus as in *4*.
2 ml 40% (w/v) polyethylene glycol (PEG; MW 6000) + 3 mM $CaCl_2$.
10% mannitol in 0.1 mM $CaCl_2$.
Osmotically stabilized salts solution (SS; see *1*).
Carbenicillin (0.4 g/ml).
Mycostatin (5 x 10^4 units/ml).

Procedure
   a. Mix the RNA and PEG solutions and store at 0°C until used.
   b. Resuspend the washed protoplasts (0.5-1.5 x 10^7) in 0.2 ml mannitol
      and mix with the PEG solution containing RNA.
   c. After *c*. 10 s dilute the suspension with 20 ml mannitol + $CaCl_2$
      with gentle shaking and incubate for 30 min at 20°C.
   d. Wash the protoplasts three times with mannitol + $CaCl_2$ and
      resuspend in 20 ml SS containing 20 µl of each of the two anti-
      biotics.
   e. Incubate in 10 ml volumes in 100 ml conical flasks at 22°C with
      continuous diffuse lighting for approximately 38 h.

*4. Assay of infected protoplasts: extraction of infectious viral RNA for
assay on local lesion host (after Wilcockson & Hull, 1974)*

Reagents required
10% mannitol.
Freshly prepared phenol mixture (0.1% p-hydroxyquinoline + 15% m-cresol
in water-saturated phenol).
Freshly prepared aqueous mixture (3.2% sodium dodecyl sulphate + 1.3%
bentonite in 0.18 M sodium phosphate buffer, pH 7.4).
Water-saturated diethyl ether.
Carborundum, F500.

Procedure
   a. Decant protoplasts into 50 ml conical centrifuge tubes (taking 0.2
      ml samples from each tube for viability assay; see *1.f.*), wash with
      mannitol and resuspend in 0.6 ml mannitol.
   b. Add 0.4 ml of the aqueous mixture and 1.01 ml of the phenol mixture
      at 0°C.
   c. Shake vigorously on a wrist action shaker for 10 min at 4°C.
   d. Centrifuge at 4500 *g* for 10 min.
   e. Separate the phases and remove excess phenol from the aqueous phase
      by washing four times with two volumes of ether, blowing off the
      remaining ether using an air line.
   f. Assay without dilution on carborundum dusted cowpeas by fingertip
      inoculation using a muslin pad.

5. Assay of infected protoplasts: indirect fluorescent antibody staining

Reagents required
10% mannitol
Acetone dried over $CaCl_2$.
Phosphate buffered saline (PBS; 1.45 M NaCl in 4.5 mM sodium phosphate buffer, pH 7.1).
Rabbit anti-CMV-W (titre 1/64 by counterimmuno-electrophoresis) previously adsorbed with freeze-dried cucumber sap to reduce non-specific staining and diluted 1 in 8 for use.
Fluorescein-labelled sheep anti-rabbit immunoglobulin (Wellcome Reagents Ltd, Beckenham, Kent).
10 mM sodium carbonate buffer in 90% glycerol, pH 8-9.

Procedure
   a. Decant protoplasts into 50 ml conical centrifuge tubes (taking 0.2 ml samples from each tube for viability assay; see *1.f.*) and wash with mannitol.
   b. Smear one drop of a suspension of infected protoplasts in mannitol on a glass microscope slide.
   c. Dry the slide rapidly in a stream of warm air.
   d. Fix the dried smear in dried acetone for 30 min.
   e. Gently rinse in distilled water to remove mannitol and equilibrate in PBS for 15 min.
   f. Cover the smear with anti-CMV serum and allow to interact with the fixed cells for 2 h at 20°C in a moist chamber.
   g. Wash slides for 30 min in PBS, drain and add fluorescein-labelled sheep anti-rabbit immunoglobulin diluted 1 in 16 with PBS to the slide.
   h. After 30 min wash slide thoroughly in fresh PBS, mount in buffered glycerol and examine under a fluorescence microscope with exciter filter KP500 and barrier filters LP520 and LP540.

FURTHER COMMENTS

Through experience obtained from the routine use of these techniques in our laboratory we have identified some of the more critical steps in the procedures. Careful attention to these steps will permit the efficient and reproducible infection of cucumber protoplasts.

   The age of the cucumber tissue is important. Fully expanded first leaves taken just before expansion of the second leaves are recommended, since first leaves which are not fully expanded are very fragile and difficult to peel, while fully expanded but senescing first leaves are tough and give unstable protoplasts.

   While growth conditions of the plants are important for infection, since they determine the ease with which the tissue is digested, the condition of the protoplasts following digestion appears to be more important. Optimally, protoplast suspensions obtained from near-complete digestion of the tissue should contain no cells and very few protoplasts misshapen by remnants of cell wall.

For protoplast infection the specific infectivity of the infecting agent is important. Routinely, infection of cucumber protoplasts with viral nucleoprotein is carried out using virus suspensions with a dilution end-point at 2 mg/ml of $0.5-1.0 \times 10^{-5}$ when assayed on cowpea. For infections using viral RNA, best results have been obtained with dilution end-points at 0.5 mg/ml of $c. 5.0 \times 10^{-3}$.

As a guide, some results obtained from typical experiments are listed in Table 1.

Table 1. Infectivities from cucumber protoplasts following inoculation with CMV or CMV-RNA

		Mean lesion number/ cowpea leaf/ 10^6 viable protoplasts	% viable protoplasts 38 h after inoculation	% viable protoplasts fluorescing*	Concentration of viable protoplasts 38 h after inoculation $(x\ 10^6/ml)$
Inoculation with CMV/PLO complexes					
Exp. 1		57.4	39.7	49.3	0.34
	2	50.2	36.8	51.2	0.33
	3	66.4	65.0	41.4	0.58
Inoculation with CMV-RNA/PEG					
Exp. 1		104.5	44.5	48.4	0.36
	2	96.3	60.4	43.4	0.57
	3	122.7	56.6	87.2	0.81

*Fluorescent antibody staining

Acknowledgements

The authors are grateful to the Leverhulme Trust for financial support

to A.J. Maule and M.I. Boulton, and to the Royal Society for provision
of controlled environment cabinets.

REFERENCES

Beier H. & Bruening G. (1976) Factors influencing the infection of
    cowpea protoplasts with cowpea mosaic virus RNA. *Virology* 72, 363-9.
Cocking E.C. (1966) An electron microscopic study of the infection of
    isolated tomato fruit protoplasts by tobacco mosaic virus. *Planta*
    68, 206-14.
Coutts R.H.A. & Wood K.R. (1976) Investigations on the infection of
    cucumber mesophyll protoplasts with cucumber mosaic virus. *Archives
    of Virology* 52, 307-13.
Coutts R.H.A. & Wood K.R. (1977a) Improved isolation and culture
    methods for cucumber mesophyll protoplasts. *Plant Science Letters* 9,
    45-51.
Coutts R.H.A. & Wood K.R. (1977b) Inoculation of leaf mesophyll
    protoplasts from a resistant and a susceptible cucumber cultivar with
    cucumber mosaic virus. *FEMS Microbiology Letters* 1, 121-4.
Dawson J.R.O., Dickerson P.E., King J.M., Sakai F., Trim A.R.H. &
    Watts J.W. (1978) Improved methods for infection of plant proto-
    plasts with viral ribonucleic acid. *Zeitschrift für Naturforschung*
    33c, 548-51.
Maule A.J., Boulton M.I. & Wood K.R. (1979a) An improved method for the
    infection of cucumber leaf protoplasts with cucumber mosaic virus.
    Phytopathologische Zeitschrift 97, 118-26.
Maule A.J., Boulton M.I., Edmunds C. & Wood K.R. (1979b) Polyethylene
    glycol mediated infection of cucumber protoplasts by cucumber mosaic
    virus and viral RNA. *Journal of General Virology* 47, 199-203.
Mühlbach H-P., Camacho-Henriquez A. & Sanger H.L. (1977) Infection of
    tomato protoplasts by ribonucleic acid of tobacco mosaic virus and by
    viroids. *Phytopathologische Zeitschrift* 90, 289-305.
Otsuki Y., Shimomura T. & Takebe I. (1972) Tobacco mosaic virus
    multiplication and expression of the N gene in necrotic responding
    tobacco varieties. *Virology* 50, 45-50.
Otsuki Y. & Takebe I. (1976) Double infection of isolated tobacco leaf
    protoplasts by two strains of tobacco mosaic virus. *Biochemistry and
    Cytology of Plant-Parasite Interactions* (Ed. by K. Tomiyama,
    J.M. Daly, I. Uritani, H. Oku & S. Ouchi), pp. 213-22. Elsevier,
    Amsterdam.
Sarkar S., Upadhya M.D. & Melchers G. (1974) A highly efficient method
    of inoculation of tobacco mesophyll protoplasts with ribonucleic acid
    of tobacco mosaic virus. *Molecular and General Genetics* 135, 1-9.
Scott H.A. (1963) Purification of cucumber mosaic virus. *Virology* 20,
    103-6.
Takebe I. (1975) The use of protoplasts in plant virology. *Annual
    Review of Phytopathology* 13, 105-25.
Takebe I. (1977) Protoplasts in the study of plant virus replication.
    Comprehensive Virology, 11 (Ed. by H.H. Fraenkel-Conrat &
    R.R. Wagner), pp. 237-83. Plenum, New York.
Takebe I. & Otsuki Y. (1969) Infection of tobacco mesophyll protoplasts

by tobacco mosaic virus. *Proceedings of the National Academy of Sciences, USA* 64, 843-8.

Wilcockson J. & Hull R. (1974) The rapid isolation of plant virus RNAs using sodium perchlorate. *Journal of General Virology* 23, 107-11.

Wood K.R., Boulton M.I. & Maule A.J. (1979) Application of protoplasts in plant virus research. *Plant Cell Cultures: Results and Perspectives* (Ed. by O. Ciferri & B. Parisi), pp. 405-10. Elsevier, Amsterdam.

Zaitlin M. & Beachy R.N. (1974) The use of protoplasts and separated cells in plant virus research. *Advances in Virus Research* 19, 1-35.

Isolation and infection of cowpea primary leaf protoplasts with tobacco necrosis virus

S.J. OLDFIELD & R.H.A. COUTTS

Department of Botany, Imperial College
of Science and Technology, London SW7 2BB

INTRODUCTION

The potential offered by protoplast systems for studying aspects of plant virus replication is well documented (Takebe, 1977). However, the uncertainty of protoplast isolation and infection has meant that this potential has not yet been fully realized. Simplification of procedures for production, isolation, infection and assay is of prime importance if plant protoplasts are to be used effectively in biochemical studies with viruses.

The method described here illustrates such a simplified approach: it employs plants with a short growth period, techniques for rapid release of protoplasts and a simple assay procedure. However, the conditions described should be investigated further and optimized for each environmental component, as indicated by reference to other similar systems (Hibi et al., 1975; Alblas & Bol, 1977). A specific biochemical means of measuring virus infection would facilitate this optimization.

ORGANISMS

Cowpea (*Vigna sinensis*) cv. Blackeye No. 5 should be used as a source of protoplasts. Tobacco necrosis virus (TNV_D) should be purified from infected French beans by a routine differential centrifugation method based on that of Salvato & Fraenkel-Conrat (1977) to give a suspension of 2-20 mg virus/ml.

METHOD

Growth of plants

Reagents required   A solution of Heller's (1953) salts, stored as three stocks, as follows (g/l).

1. KCl, 7.5; $CaCl_2$, 0.75; $NaNO_3$, 6.0; $MgSO_4.7H_2O$, 2.5; $NaH_2PO_4.2H_2O$, 1.25.
2. $FeCl_3.6H_2O$, 1.0.
3. $MnSO_4.4H_2O$, 0.1; $ZnSO_4.7H_2O$, 1.0; H_3BO_3, 1.0; KI, 0.01; $CuSO_4.5H_2O$, 0.03; $NiCl_2.6H_2O$, 0.03.

To make 1 l of Heller's salts solution mix together 100 ml of stock 1, 1.0 ml of stock 2 and 1.0 ml of stock 3, then make up to 1 l with distilled water.

Procedure   For optimum protoplast yield and viability germinate cowpea in moistened Vermiculite in a growth cabinet at 25°C with a 16 h, 10 000 lx photoperiod (from fluorescent and incandescent sources). After 4 days transfer the plants to bottles filled with a solution of Heller's salts and maintain under the same conditions for 4–8 days.

Isolation of protoplasts

Reagents required
1. Sterilization solutions: 70% ethanol; 5% solution of Domestos bleach (10% available chlorine).
2. Plasmolysing solution: a 0.5 M solution of mannitol (91.1 g/l) sterilized at 120°C for 15 min.
3. Enzyme solution (Beier & Bruening, 1976) prepared and sterilized on day of use, consisting of (g/100 ml): mannitol, 9.11; bovine serum albumen, 0.1; Macerozyme R10 (Yakult Manufacturing Co., Japan), 0.1; cellulase Onozuka R10 (Yakult Manufacturing Co.), 0.5; pH adjusted to 5.5.

Prodecure   Select fully expanded primary leaves from the cowpea plants just prior to emergence of the first trifoliate leaf (Hibi *et al.*, 1975). Surface sterilize these by washing in 70% ethanol for 5 s, immersing in a 5% solution of Domestos for 15 min and finally washing three times in sterile distilled water. The use of large casserole dishes facilitates these aseptic washing and handling procedures.

Detach the lower epidermis from the aseptic leaves, using jeweller's forceps, and float the leaves, peeled side down, in sterile Petri dishes (7.5 cm diam.) containing plasmolysing solution. Approximately three leaves (1.5 g) should be placed in each dish. After at least 60 min replace the plasmolysing solution in each dish with 25 ml freshly prepared enzyme solution. Seal the dishes with Parafilm and incubate in darkness in a shaking water bath at 30°C and 25 excursions/min.

After incubating for 3.5 h filter the protoplast suspension through sterile nylon bolting cloth (mesh size 77 μm) into sterile centrifuge tubes. The material can best be transferred using sterile disposable wide-bore Pasteur pipettes. Sediment in a bench centrifuge at 200 *g* for 3 min. Decant the enzyme solution and wash the protoplasts three

times by repeated suspension in plasmolysing solution followed by
sedimentation as above.

Resuspend the protoplasts in plasmolysing solution and determine
their concentration by counting in a Fuchs-Rosenthal haemocytometer.
Adjust the concentration to 3-5 x 10^6 protoplasts/tube.

Infection and culture of protoplasts

Reagents required
  1. Phosphate buffered mannitol solution (PBM; Wieringa-Brants *et al.*,
     1978): 0.2 M phosphate buffer (pH 5.6) in 0.7 M mannitol,
     sterilized at 120°C for 15 min.
  2. Protoplast wash solution (MC): mannitol (91.1 g/l); $CaCl_2.2H_2O$
     (14.7 mg/l); sterilized at 120°C for 15 min.
  3. Virus incubation medium (VIM; Aoki & Takebe, 1969; mg/l): KH_2PO_4,
     13.6; KNO_3, 50.5; $MgSO_4$, 123.0; $CaCl_2.2H_2O$, 740.0; KI, 0.166;
     $CuSO_4.5H_2O$, 0.249; mannitol, 91 100 adjusted to pH 5.4 with 0.5
     N KOH and sterilized at 120°C for 15 min.
  4. Antibiotic stock solutions (Watts & King, 1973): Nystatin (75
     units/1 ml VIM), stored at 20°C; chloramphenicol (0.05 g/25 ml
     VIM), prepared on the day of use and filter sterilized.

Procedure   Pre-incubate poly-L-ornithine (mol. wt. 122 000; stock
solution 200 µg/ml) with TNV_D (2-20 mg/ml) and PBM in 0.7 M mannitol
to give the required final concentrations (e.g. 2 µg/ml poly-L-
ornithine, 5 µg/ml TNV_D, and 0.01 M PBM) and maintain for 5 min at
25°C with shaking (Alblas & Bol, 1977; Wieringa-Brants *et al.*, 1978).

Sediment the protoplasts by centrifugation at 200 *g* for 3 min and
immediately resuspend in 10 ml of the pre-incubated mixture contained
in 100 ml Erlenmeyer flasks.   Incubate the suspension at 0°C for 15 min
on a reciprocal shaker at 25 excursions/min.

Terminate the infection procedure by sedimenting the protoplasts as
above and washing three times with MC.   Resuspend the washed protoplasts in VIM containing Nystatin stock solution (0.15 ml/100 ml VIM)
and chloramphenicol stock solution (10 ml/90 ml VIM) to give 2.5 x 10^5
protoplasts/ml.   Incubate 1.0 ml aliquots of the suspension in
scintillation vials at 22°C with continuous light (3000 lx) until
harvesting.   Freeze the harvested samples at -20°C and store until all
the samples can be assayed together.

Rapid virus assay

Thaw the samples and dialyze 1 ml aliquots overnight against distilled
water at 4°C.   Ensure protoplast disruption by sonication and dilute
the samples (1 in 10) in order to moderate lesion counts.   Assay the
infective virus content using a local lesion assay on the primary
leaves of 10- to 12-day old glasshouse grown cowpea plants (Coutts,
1978).

DISCUSSION

Cowpea mesophyll protoplasts were chosen for studying viral replication
at the biochemical level because of the good experimental control which
such a system affords. The growth cycle is shorter than that of
tobacco, allowing more physiological control of the material throughout
the culture procedure, a feature which is essential if results are to
be reproducible. In addition, the wide susceptibility to different
viruses and the convenient method of rapid protoplast isolation and
virus infection make cowpea an ideal experimental tool.

   The method described here is essentially a single step protoplast
isolation (Hibi et al., 1975) with a direct virus infection (Kubo et
al., 1975) incorporating an osmotic "step-up" (Okuno & Furusawa, 1978).
This procedure is rapid and can easily be controlled and optimized.
The vial incubation and assay procedure provide a ready assessment of
virus replication, although a more rapid method of assay (e.g. fluorescent antibody staining) would be an advantage and is being developed.

   One duplicated experiment, using cowpea protoplasts isolated and
infected as above, illustrates how the concentration of poly-L-
ornithine, used during a routine infection, has been provisionally
optimized (Table 1). Optimization of other aspects of isolation,
infection and assay procedures is now well advanced, and when complete
will provide a basis for sophisticated experiments to investigate the
more complex aspects of virus synthesis. For example, interference
from associated, possibly defective, virus particles such as satellite
virus can be studied through replicase activity in a synchronous
fashion; this was not previously possible with whole leaf systems.

Table 1. The effects of different concentrations of poly-L-ornithine
on the infection of cowpea protoplasts by tobacco necrosis virus in
duplicated experiments. Results are expressed as the mean number
of lesions/cowpea leaf in local lesion assays.

| Poly-L-ornithine concentration µg/ml | Hours after inoculation | | | | | |
| | 0 | | 24 | | 48 | |
	exp.1	exp.2	exp.1	exp.2	exp.1	exp.2
0.5	0.3	0.3	0.3	0.3	1.0	4.1
1.0	0.2	0.3	1.8	0.6	0.6	1.1
2.0	0.3	0.2	39.6	19.1	38.5	135.1
4.0	0.4	0.3	46.3	51.4	120.6	118.2
8.0	0.3	0.3	2.4	1.0	16.4	24.8
12.0	0.3	0.4	0.2	0.4	3.8	5.6
16.0	0.3	0.2	0.2	0.3	0.8	0.3

The isolation procedure described here routinely yields 3.5 x 10^6 protoplasts/g leaf material with over 95% viability as assessed by phenosafranine or Evan's blue staining (Widholm, 1972). Uninfected protoplasts were used in labelling studies and incorporation monitored as previously (Coutts *et al.*, 1975). After an initial lag period both [^{14}C] leucine and [^{3}H] uridine were incorporated throughout the 24 h period tested, illustrating that the metabolic activity of the protoplasts is well suited to studies of protein and nucleic acid synthesis.

REFERENCES

Alblas F. & Bol J.F. (1977) Factors influencing the infection of cowpea mesophyll protoplasts by alfalfa mosaic virus. *Journal of General Virology* 36, 175-87.

Aoki S. & Takebe I. (1969) Infection of tobacco mesophyll protoplasts by tobacco mosaic virus ribonucleic acid. *Virology* 39, 439-48.

Beier H. & Bruening G. (1976) Factors influencing the infection of cowpea protoplasts by cowpea mosaic virus RNA. *Virology* 72, 363-9.

Coutts R.H.A. (1978) Suppression of virus induced local lesions in plasmolysed leaf tissue. *Plant Science Letters* 12, 77-85.

Coutts R.H.A., Barnett A. & Wood K.R. (1975) Ribosomal RNA metabolism in cucumber leaf mesophyll protoplasts. *Nucleic Acids Research* 2, 1111-21.

Heller R. (1953) Recherches sur la nutrition minérale des tissus végétaux cultivés *in vitro*. *Annales des sciences naturelles* (Botanique et Biologie Végétale) 14, 1-223.

Hibi T., Rejelman G. & Van Kammen A. (1975) Infection of cowpea mesophyll protoplasts with cowpea mosaic virus. *Virology* 64, 308-18.

Kubo S., Harrison B.D., Robinson D.J. & Mayo M.A. (1975) Tobacco rattle virus in tobacco mesophyll protoplasts: infection and virus multiplication. *Journal of General Virology* 27, 293-304.

Okuno T. & Furusawa U. (1978) The use of osmotic shock for the inoculation of barley protoplasts with brome mosaic virus. *Journal of General Virology* 39, 187-90.

Salvato M.S. & Fraenkel-Conrat H. (1977) Translation of tobacco necrosis virus and its satellite in a cell-free wheat germ system. *Proceedings of the National Academy of Sciences, USA* 74, 2288-92.

Takebe I. (1977) Protoplasts in the study of plant virus replication. *Comprehensive Virology, Vol. II* (Ed. by H. Fraenkel-Conrat & R.R. Wagner), pp. 237-83. Plenum, New York.

Watts J.W. & King J.M. (1973) The use of antibiotics in the culture of non-sterile plant protoplasts. *Planta* 113, 271-7.

Widholm J.M. (1972) The use of fluorescein diacetate and phenosafranine for determining viability of cultured plant cells. *Stain Technology* 47, 189-93.

Wieringa-Brants D.H., Timmer F.A. & Rouweler M.H.C. (1978) Infection of cowpea mesophyll protoplasts by tobacco mosaic virus and tobacco necrosis virus. *Netherlands Journal of Plant Pathology* 84, 239-40.

Methodological aspects of virus infection and replication in plant protoplasts

A.C. CASSELLS* & F.M. COCKER†

*Department of Botany, University College, Cork
†Department of Horticulture, Wye College
(University of London), Wye, Kent TN25 5AH

INTRODUCTION

The polycation infection procedure developed by Takebe and co-workers has found wide acceptance among those studying virus replication in plant protoplasts (Takebe, 1975). However, in our earlier investigation of tobacco mosaic virus (TMV) replication in tomato protoplasts (Cassells & Barlass, 1978a) we found the polycation procedure to be toxic. This, together with uncertainty surrounding the mechanism (Burgess et al., 1973) led us to investigate alternative infection procedures; a modification of the polyethylene glycol (PEG) procedure (Cassells & Barlass, 1978b) is outlined here.

MATERIALS AND METHODS

Protoplast isolation

Grow *Nicotiana tabacum* cv. xanthi-nc plants in the glasshouse in 12.7 cm pots containing Irish Peat Moss supplied with a complete balanced fertilizer (Bio P Base; Pan Britannica Industries, Hertfordshire). Shade the plants in summer and provide with supplementary light in winter (HPLR 400 W lamps; Philips Ltd, London) to maintain approximately autumn light intensities (c. 5–10 MJ/m^2/day) throughout the year with a 15 h daylength. Grow plants to the 10–20 leaf stage and use uniform leaves (lamina length c. 22 cm) for protoplast isolation. Isolate protoplasts by floating peeled leaves on a mixture of 1% w/v Onozuka R10 cellulase and 0.05% w/v Macerozyme R10 pectinase (Yakult Manufacturing Co., Japan) in 0.66 M mannitol at pH 5.8 for 1–2 h at 25°C in the dark. Swirl gently to obtain maximum release of protoplasts. Transfer the protoplasts to 10 ml conical tubes using a wide bore Pasteur pipette and allow to settle under gravity. Decant the supernatant, replace with fresh 0.66 M mannitol solution and allow the protoplasts to settle again. Repeat this washing twice. Dilute an aliquot of the protoplast suspension to give approximately 10^5 protoplasts per ml, then make an accurate count of numbers using a modified Fuchs-Rosenthal haemocytometer (Hawksley, Lancing, Sussex).

Ingram D.S. & Helgeson J.P. (1980) *Tissue Culture Methods for Plant Pathologists*

Protoplast viability

Determine protoplast viability by staining with fluorescein diacetate
(FDA; Widholm, 1972). Dissolve 2-3 mg of FDA in 0.5 ml of acetone and
then add 10 ml of 0.66 M mannitol at pH 5.8. Mix together one drop of
protoplast suspension and one drop of stain on a microscope slide and
leave in the dark for 5-10 min. View the slides with a fluorescence
microscope equipped with a BG 38 suppressor filter, an FITC 5 Balzer
exciter filter and OG 515 and OG 530 barrier filters. Live protoplasts
fluoresce yellow-green. Determine the mean percentage of viable
protoplasts in 10 fields.

Infection procedure

To a suspension of 10^6 protoplasts in 0.4 ml 0.66 M mannitol at pH 5.8
add 0.1 ml of 83.3 mM PEG 6000 in 0.66 M mannitol, and 10 µl TMV (12 mg/
ml) in 0.5 ml of 5 mM phosphate-buffered 0.66 M mannitol at pH 6.5.
Incubate at 4°C for 20 min. After incubation, add 0.3 ml of 83.3 mM PEG
6000 in 0.66 M mannitol and 0.7 ml of 3.5 mM phosphate buffer at pH 6.5
containing 11 mM $CaCl_2$ in 0.66 M mannitol. Incubate for 1 h at 25°C in
the dark. Allow the protoplasts to settle under gravity, decant the
supernatant and resuspend the protoplasts in 50 mM $CaCl_2$ in 0.66 M
mannitol at pH 10.5. Allow the protoplasts to settle under gravity,
resuspend the pellet in liquid medium (pH 6.5; Nagata & Takebe, 1970)
and again allow to settle under gravity. Resuspend the protoplasts
(10^6) in 5 ml of liquid medium (Nagata & Takebe, 1970) and incubate at
22°C with a 16 h photoperiod (8.6 W/m^2).

Assay of infectious virus production

After 24 or 48 h (as appropriate) collect the protoplasts by gravity
sedimentation and adjust the suspension to contain 10^6 protoplasts in
0.4 ml of medium. Remove 0.15 ml of this suspension and stain an
aliquot with FDA (as above) to determine protoplast viability; stain a
further aliquot with fluorescent antibody to determine the percentage of
infected protoplasts (see below). Dilute the remaining 0.25 ml of the
protoplast suspension to 1 ml with 0.1 M sodium phosphate buffer at pH
7.0. Dialyse these protoplasts, together with 1 ml volumes of the
supernatant from the infected protoplast suspension, overnight at 4°C in
phosphate buffered saline (PBS: 0.01 M sodium phosphate buffer at pH 7.0
containing 0.15 M sodium chloride).

After dialysis and before inoculation of *N. tabacum* cv. xanthi-nc
test plants, examine the samples under the electron microscope following
negative staining with uranyl acetate (Noordam, 1973) to estimate virus
concentration. On the basis of this examination dilute the protoplast
extracts to give approximately equivalent particle numbers to a 1:1000
dilution of the standard virus solution (12 mg/ml TMV in 50 mM sodium
phosphate buffer at pH 7.0).

Add 1 mg celite to 0.3 ml of the samples and inoculate half leaves
of xanthi-nc plants. Inoculate the opposite half leaves with standard

virus preparation diluted as above. Assay each sample at least three times. Allow local lesions to develop in the growthroom at 25°C with a 16 h photoperiod (8.6 W/m^2).

Determination of percentage infection using fluorescent antibody staining

Fix the protoplasts pellet (*c.* 3 x 10^5 protoplasts in 0.15 ml) by adding 2-3 ml of fresh Carnoy's reagent (absolute alcohol:chloroform: glacial acetic acid, 6:3:1) at 4°C. Decant the Carnoy's reagent after 10 min and add 10 ml of acetone. Leave for 2 days until decolourization is complete. Centrifuge down the flocculence which contains the protoplasts at 1500 *g* for 10 min. Add distilled water to the pellet to wash free of mannitol. Repeat the wash and centrifugation steps. Coat two microscope slides with Mayer's albumen. Place a drop of protoplast suspension on each of the slides, smear and dry quickly in a stream of warm air. Pre-incubate one of the replicate slides (the control) with unconjugated antibody in a humid container for 2 h at 37°C. Wash for 2 h with PBS at 25°C. Coat both slides with a thin layer of FITC conjugated antiserum (for preparation see Cassells & Gatenby, 1975) and incubate in the dark at 37°C in a humid chamber for 2 h. Wash the slides overnight in the dark at 4°C in PBS at pH 7.0. View the slides with a fluorescence microscope using the filter system described above for FDA.

Efficiency of infection procedure

Typically, using the original PEG inoculation procedure (Cassells & Barlass, 1978b), protoplast viability 48 h after infection was 60-70%, with virus antigen detectable in *c.* 60% of the protoplasts and a mean of 1.9 x 10^6 progeny particles produced per infected protoplast. Using the modified PEG procedure described here virus inoculum concentration was reduced 5-fold without affecting progeny virus production or percentage protoplast infection.

DISCUSSION

The PEG infection procedure differs in at least one major aspect from the polycation procedure (Takebe, 1975). In the latter, relatively large numbers of virus particles are seen in vesicles at the plasmalemma and in the cytoplasm in the early stages of infection. No vesicle formation has been detected in the early stages of the PEG infection procedure (Cassells & Cocker, unpublished data). Considerable contro-versy surrounds the mechanism of infection in the polycation procedure. Some workers believe that virus particles taken up by pinocytosis initiate infection (Takebe, 1975) while others (Burgess *et al.*, 1973; Kassanis *et al.*, 1977) consider that virus entering through polycation-induced membrane lesions may also initiate infection. The polycation and PEG procedures provide systems for the study of the processes in plant virus replication. Their further investigation and comparison may provide an insight into the infection process itself, with the caveat that the cell wall may play a significant role in the initiation of infection in the intact plant.

REFERENCES

Burgess C., Motoyoshi F. & Fleming E.N. (1973) The mechanism of infection of plant protoplasts by viruses. *Planta* 112, 323-32.

Cassells A.C. & Barlass M. (1978a) A method for the isolation of stable mesophyll protoplasts throughout the year under standard conditions. *Physiologia Plantarum* 42, 236-42.

Cassells A.C. & Barlass M. (1978b) The initiation of TMV infection in isolated protoplasts by polyethylene glycol. *Virology* 87, 459-62.

Cassells A.C. & Gatenby A.A. (1975) The use of lissamine rhodamine B conjugated antibody for the detection of tobacco mosaic virus antigen in tomato protoplasts. *Zeitschrift für Naturforschung* 30c, 696-7.

Kassanis B., White R.F., Turner R.H. & Woods R.D. (1977) The mechanism of virus entry during infection of tobacco protoplasts with TMV. *Phytopathologische Zeitschrift* 88, 215-28.

Nagata T. & Takebe I. (1970) Cell wall regeneration and cell division in isolated tobacco mesophyll protoplasts. *Planta* 99, 12-20.

Noordam D. (1973) *Identification of Plant Viruses*. Centre for Agricultural Publishing and Documentation, Wageningen, Netherlands.

Takebe I. (1975) The use of protoplasts in plant virology. *Annual Review of Phytopathology* 13, 105-25.

Widholm J. (1972) The use of fluorescein diacetate and phenosafranine for determining viability of cultured plant cells. *Stain Technology* 47, 189-94.

Infection of plant protoplasts with two or more viruses

J.W. WATTS, J.R.O. DAWSON & J.M. KING
John Innes Institute, Colney Lane,
Norwich NR4 7UH

INTRODUCTION

Plant protoplasts can be efficiently and synchronously infected with viruses. This process is followed by a single round of viral multiplication without cell lysis or cross infection. The attractiveness of the system has encouraged many workers to use protoplasts, but since great difficulties may be experienced both in producing and infecting them, and since many aspects of virology can be investigated with intact plants, it is probably better to use them only when there is no alternative. Protoplasts can be infected with several viruses simultaneously, and they are therefore ideally suited to the study of mixed infections at the cellular level, something that is almost impossible with intact plants. Relatively little work on these lines has been published, although studies of protoplasts infected with different strains of the same virus and with different viruses have been reported (Dawson et al., 1975; Otsuki & Takebe, 1976, 1978; Barker & Harrison, 1978; Dawson & Watts, 1979).

The procedures outlined below deal with the simultaneous inoculation of tobacco protoplasts with cowpea chlorotic mottle virus (CCMV) and brome mosaic virus (BMV).

METHODS

Growth of plants

Good plants will produce protoplasts of great stability. The following details refer specifically to tobacco but are applicable with modification to other species.

Place about 20 seeds on a moist filter paper in a Petri dish, tape with Parafilm and incubate at 25°C; good tobacco seed germinates in 2 days. Fill ten 50 cm^3 disposable plastic pots with a peat based potting compost, level the surface and place one germinated seed in each; sift a thin layer of compost onto the seeds. Water gently with a wash-bottle

and incubate at 22-25°C with a 14 h day. Plants emerge within 2-3 days.
Plants require 4-7 x 10^3 lx for satisfactory growth. We use an array of
10 x 60 W Philips white fluorescent tubes with 6 x 60 W tungsten lamps
to illuminate an area about 1.6 x 2 m from a height of about 1.5 m.
Humidity should be about 60-70%.

When the plants are about 3 weeks old (3-4 cm diam.) transfer to
compost in 175 mm pots. Avoiding unnecessary root disturbance, simply
drop the compost block containing the plant into a hole in the new
compost. Grow on as before but after 2 weeks begin to water with
Phostrogen solution (0.5 g/l; Phostrogen Ltd, Corwen, Clwyd, UK) two to
three times a week. Avoid over-watering. Xanthi is ready to use when
about 40 cm tall, White Burley when it is 6-7 weeks old and has a
rosette form with three large pointed leaves (Watts *et al.*, 1974).

Preparation of protoplasts

Solutions required   Use distilled water of good quality (Analar if
in doubt or fractionally distil).
  1. 0.7 M mannitol.
  2. 0.5 g Macerozyme R10 (Yakult Manufacturing Co., Japan), 0.5 g
     potassium dextran sulphate in 100 ml 0.7 M mannitol, adjusted to
     pH 5.5 with phosphoric acid.
  3. 0.5 g Onozuka R10 cellulase (Yakult Manufacturing Co., Japan) in
     50 ml 0.7 M mannitol, adjusted to pH 5.5 with NaOH.

Procedure   Select a suitable leaf, about 75% expanded; it may help
peeling if the leaf is allowed to dry out on the bench for 30 min.
Remove the lower epidermis using forceps. Almost any type of forceps
can be used; flat (spade) ended or curved are popular but straight
forceps are as good in trained hands; the main requirement is accurate
mating of the tips. Cut the peeled leaf into pieces (2 cm^2) and
transfer to 100 ml 0.7 M mannitol in a 250 ml Erlenmeyer flask. When
all the leaf pieces have been collected, decant the 0.7 M mannitol and
replace with 30 ml Macerozyme solution. The initial wash and pre-
plasmolysis of peeled leaf material is essential with some species,
helping to remove toxic substances released by damaged tissue. Vacuum
infiltrate the leaf pieces for 30 s at 10 mmHg. Shake on a reciprocating water bath at 25°C, stroke 2.5 cm, frequency 2 Hz, for 5 min.
Discard the solution with the debris from damaged cells and replace with
35 ml Macerozyme solution; shake as before but for 20 min. Collect the
cells by filtering through 1 mm nylon mesh. Return the leaf pieces to
the Erlenmeyer flask and add the rest of the Macerozyme solution; shake
for a further 30 min and collect the cell suspension as before.

The standard conditions for centrifuging cells or protoplasts are 600
rev/min for 2 min in a bench centrifuge. If necessary, use a cheap
drill-speed controller (thyristor type) to slow down the centrifuge.

Wash the cells once with 0.7 M mannitol, resuspend in the Onozuka
(cellulase) solution in the original flask and shake on a reciprocating
water bath at 35°C, stroke 2.5 cm, frequency 1 Hz. Protoplasts normally

form within 30-40 min. After 1 h, if all cells appear rounded and
without walls, filter the protoplasts through two thicknesses of
cheesecloth and wash with 0.7 M mannitol until the supernatant fluid is
only slightly coloured. If the supernatant fluid remains turbid after
successive washes, abandon the experiment as the protoplasts are too
unstable.

Inoculation with virus

The procedure uses a polycation which increases the susceptibility of
the plasmalemma and, in the case of RNA and negatively charged viruses,
causes aggregation and modifies surface charge so that the infective
particles can attach to the protoplasts.

Solutions required (Motoyoshi et al., *1973, 1974)*
   1. 0.7 M mannitol.
   2. 0.1 M sodium citrate buffer, pH 5.2.
   3. Stock solution of poly-L-ornithine (mol. wt. 120 000) in water
      (1 mg/ml; Sigma Chemical Co., London).
   4. Virus suspensions (1-10 mg/ml).

Procedure   While the protoplasts are being washed, make up the CCMV
inoculum with the composition: 0.7 M mannitol (90 ml); citrate buffer
(10 ml); poly-L-ornithine solution (0.1 ml); CCMV (100 µg). Mix and
allow to pre-incubate for 10 min.

   Spin down the protoplasts from their final wash, and immediately
resuspend in the inoculum. Use 10 ml inoculum for a maximum of 10^6
protoplasts. Leave the mixture, with occasional mixing, for 10 min.
Wash once with 0.7 M mannitol.

   The inoculum for the second virus is prepared during the first
inoculation and has the composition: 0.7 M mannitol (90 ml); citrate
buffer (10 ml); poly-L-ornithine solution (0.1 ml); BMV (500 µg). Mix
and allow to pre-incubate for 10 min.

   After the first inoculation resuspend the protoplasts as before
immediately after they have been spun down. Leave for 10 min with
occasional mixing. Wash three times with 0.7 M mannitol and resuspend
in culture medium to give 1-2 x 10^5 protoplasts/ml. 50 ml Erlenmeyers
make good culture vessels; use 10 ml protoplast suspension per flask.

Culture medium   We use a simple medium with the following composition:
mannitol (0.7 M); KH_2PO_4 (0.2 mM); KNO_3 (1.0 mM); $MgSO_4$ (1.0 mM); $CaCl_2$
(10 mM); KI (1.0 µM); $CuSO_4$ (0.01 µM). No energy source is added.
Antibiotics may be added but are not essential if all glassware and
solutions are sterile. Penicillins are relatively non-toxic; e.g.
Carbenicillin (Beecham Research Laboratories, Brentford) and Cephalori-
din (Glaxo Laboratories Ltd), each at 100 µg/ml. Gentamicin (5 µg/ml;
Flow Laboratories, Irvine, Scotland) is a stable alternative but very
much more toxic. Fungi can be partially controlled by Mycostatin (5 µg/
ml; E.R. Squibb & Sons Ltd, London), but this substance decomposes so

rapidly that it must be replaced daily. Cultures are maintained at 25°C
in continuous white light (Philips white fluorescent, 500 lx).

Harvesting protoplasts

It is particularly important to use a biological assay as well as
physical or immunological assays when dealing with mixed infections.
Mixed infections with BMV and CCMV normally produce large numbers of
non-infectious CCMV particles which would be mistaken for normal virus
particles by physical and immunological assays.

Procedure   Transfer the cultures to 15 ml round bottomed tubes, wash
three times with 0.7 M mannitol and resuspend in 1 ml 0.7 M mannitol.
Place 50 µl drops on slides that have been subbed with glycerine albumen,
allow to spread and then dry in a stream of warm air. Spin down the
rest of the protoplasts and resuspend in 0.02 M acetate buffer at pH 4.8
(1-2 ml). Homogenize in a hand-operated or loosely fitting motor-driven
Potter homogenizer, spin down debris at 2000 rev/min for 10 min and
collect the supernatant fluid.

Fluorescent antibody staining   Fluorescent antibody may be prepared
by any conventional method from specific anti-virus rabbit globulins.
The slide with dried protoplasts is fixed for 15 min in ethanol and
washed for 15 min in phosphate-buffered saline (0.45 g KH_2PO_4, 0.95 g
Na_2HPO_4 and 8.5 g NaCl, made up to 1000 ml with water and adjusted to
pH 7.0). Except for a small area containing protoplasts, the slide is
dried with filter paper, and a drop of fluorescent antibody solution is
layered over the protoplasts. The slide is incubated in a moist
atmosphere in the dark at 30°C for 1 h and then washed for 1 h with
phosphate buffered saline. The protoplasts are mounted in buffered
saline and examined under the fluorescence microscope; infected cells
have a bright green fluorescence.

Assay on plants   The clarified homogenate may be rubbed onto leaves
of *Chenopodium hybridum* for local lesion assay. Since BMV gives
considerably larger lesions than does CCMV, a differential count can be
made after 3-4 days. The homogenate may be rubbed onto leaves of barley
and cowpea, which provide much more sensitive indicators for BMV and
CCMV respectively, producing distinct mottle symptoms on young leaves
after 1-2 weeks.

REFERENCES

Barker H. & Harrison B.D. (1978) Double infection, interference and
    superinfection in protoplasts exposed to two strains of raspberry
    ringspot virus. *Journal of General Virology* 40, 647-58.
Dawson J.R.O., Motoyoshi F., Watts J.W. & Bancroft J.B. (1975)
    Production of RNA and coat protein of a wild-type isolate and a
    temperature-sensitive mutant of cowpea chlorotic mottle virus in
    cowpea leaves and tobacco protoplasts. *Journal of General Virology*
    29, 99-107.

Dawson J.R.O. & Watts J.W. (1979) Analysis of the products of mixed
    infection of tobacco protoplasts with two strains of cowpea chlorotic
    mottle virus. *Journal of General Virology* 45, 133-7.
Motoyoshi F., Bancroft J.B., Watts J.W. & Burgess J. (1973) The
    infection of tobacco protoplasts with cowpea chlorotic mottle virus
    and its RNA. *Journal of General Virology* 20, 177-93.
Motoyoshi F., Watts J.W. & Bancroft J.B. (1974) Factors influencing the
    infection of tobacco protoplasts by cowpea chlorotic mottle virus.
    Journal of General Virology 25, 245-56.
Otsuki Y. & Takebe I. (1976) Double infection of isolated tobacco
    mesophyll protoplasts by unrelated plant viruses. *Journal of General
    Virology* 30, 309-16.
Otsuki Y. & Takebe I. (1978) Production of mixedly coated particles in
    tobacco mesophyll protoplasts doubly infected by strains of tobacco
    mosaic virus. *Virology* 84, 162-71.
Watts J.W., Motoyoshi F. & King J.M. (1974) Problems associated with
    the production of stable protoplasts of cells of tobacco mesophyll.
    Annals of Botany 38, 667-71.

Superinfection of mesophyll protoplasts with viruses

H. BARKER

Scottish Horticultural Research Institute,
Invergowrie, Dundee DD2 5DA

INTRODUCTION

Some aspects of virus behaviour that are difficult to examine in intact
plants are more easily studied using isolated mesophyll protoplasts. One
such aspect is the interaction of two viruses in the same host cell.
Dawson et al. (1975), Watts et al. (1975) and Otsuki & Takebe (1976a,
1976b) have used the technique of double infection of isolated tobacco
mesophyll protoplasts to investigate the interactions between related
and unrelated viruses. I have used an alternative approach to study
the interaction of viruses within a cell. This is to prepare protoplasts from systemically infected leaves of Nicotiana benthamiana
plants and to inoculate them with a second virus. One particular
advantage of this method is that one of the interacting viruses can
have replicated extensively before the second is inoculated. There are
also difficulties with this method, as infected plants can sometimes
yield protoplasts which are difficult to manipulate and infect, or
which rapidly die in culture. N. benthamiana has been a suitable
source of protoplasts for the work on superinfection, but successful
extension of the method to other virus systems will depend on a
suitable host being available.

The superinfection technique was first used to study the interaction
of raspberry ringspot virus (RRV) and the CAM strain of tobacco rattle
virus (TRV-CAM; Barker & Harrison, 1977a, 1977b), and more recently it
has proved of value in work on the interaction of two strains of RRV
(Barker & Harrison, 1978). The methods used for investigating double
infection with RRV and TRV are given below.

METHODS

Preparation of N. benthamiana *protoplasts*

Plants of *N. benthamiana* should be grown in 85 x 85 mm cross-section
pots containing a peat-sand mixture. For 2-3 weeks before isolating
protoplasts keep plants in a controlled environment, namely 16 h at

24°C with illumination of 5000 lx alternating with 8 h at 20°C in darkness. Use the youngest fully expanded leaves on 45- to 70-day old plants for protoplast isolation. All manipulations should be done in sterilized glassware, and that used for protoplast inoculation and culture should be treated with Repelcote (Hopkin & Williams Ltd) to give a siliconized surface; siliconization improves protoplast survival and aids protoplast manipulation. Use autoclaved 0.7 M mannitol as the basic medium in all stages of protoplast isolation, inoculation and culture. The mannitol solution should be made with glass redistilled water.

Isolate protoplasts by sequential enzymatic digestion of the leaf tissue using pectinase followed by cellulase (Takebe et al., 1968) as follows. Remove the lower epidermis with fine forceps before vacuum infiltrating small pieces of leaf tissue for 2 min in a pectinase (Macerozyme R10; Yakult Manufacturing Co., Japan) at 0.25% with 0.5% potassium dextran sulphate (Meito Sangyo Co. Ltd, Japan) in 0.7 M mannitol at pH 5.8. Incubate the leaf tissue at 25°C, shaking the samples at 120 excursions/min. Use three changes of pectinase, with incubation times of c. 2, 5 and 40 min respectively. The cells released during the third incubation are predominantly from the palisade mesophyll; filter them through muslin and centrifuge at 80 g for 100 s. Resuspend the pellets in 2% cellulase (Onozuka R10; Yakult Manufacturing Co.) in 0.7 M mannitol at pH 5.2, and incubate for 45 min at 35°C, shaking at 80 excursions/min. Filter the protoplasts through two layers of muslin and wash three times in fresh mannitol solution.

N. benthamiana plants infected with TRV-CAM are almost symptomless and are good sources of stable protoplasts. Leaves with severe disease symptoms from RRV-infected plants usually give low yields of proto-plasts, most of which die soon after isolation. However, in the later stages of infection, RRV-infected N. benthamiana plants produce leaves which are almost symptomless, but contain the virus (recovered leaves). Numerous stable protoplasts can be isolated from such recovered leaves. Plants infected with a severe strain of RRV (RRV-E) take 7-10 days longer after inoculation to produce suitable recovered leaves than do those infected with a less severe strain (RRV-S).

Superinfection of N. benthamiana *protoplasts*

Protoplasts should be inoculated by the indirect method (Takebe & Otsuki, 1969). A double strength inoculum should be incubated at 25°C for 10 min before being mixed with an equal volume of freshly resus-pended protoplasts and further incubated for 10 min. Use a standard inoculation mixture containing a final concentration of virus at 1 µg/ml, poly-L-ornithine (type 1C, mol. wt. 120 000; Sigma Chemical Co., London) at 1 µg/ml and $5-8 \times 10^4$ protoplasts per ml, in 0.025 M potassium phosphate buffer (pH 6.0) containing 0.7 M mannitol. After inoculation wash the protoplasts three times in 0.7 M mannitol containing 0.1 mM calcium chloride.

Keep inoculated protoplasts in incubation medium (Kubo *et al.*, 1975) for up to 3 days at 22°C, with continuous illumination of 3000 lx. After incubation treat the protoplasts with fluorescent antibody to particles of RRV or TRV-CAM. Fluorescein-conjugated antibody to virus particles should be prepared as described by Otsuki & Takebe (1969). Dry the protoplast samples quickly in a stream of warm air onto micro-scope slides which have been coated with Mayer's albumen. Fix samples with 95% ethanol for 15 min, then wash with 0.01 M phosphate buffer (pH 7.0) containing 0.85% sodium chloride (PBS) for 15 min before staining with a suitable dilution of conjugated antibody for 1 h at 37°C. After staining, wash samples in PBS for 30 min before mounting in PBS containing 40% (v/v) glycerol. Examine samples in a microscope equipped with a suitable UV optical system and assess the percentage of protoplasts which become stained (Barker & Harrison, 1977a).

Interaction between RRV and TRV

The superinfection technique described above has been used to produce important evidence on the nature of the interaction *in vivo* between particles of RRV and TRV-CAM that results in the formation of aggre-gates containing particles of both viruses (Barker & Harrison, 1977a, 1977b; Harrison *et al.*, 1977). One consequence of this phenomenon was that treatment with fluorescent antibody to RRV particles revealed

Table 1. Production of antigen aggregates by raspberry ringspot virus on superinfection of protoplasts from singly infected *Nicotiana benthamiana* plants (Data from Barker & Harrison, 1977b).

Primary infection	% protoplasts stained with homologous fluorescent antibody imme-diately after isolation	Virus used for superinfection	% protoplasts stained with fluorescent anti-body 3 days after superinfection	
			RRV-E antibody*	TRV antibody
RRV-E	99[†]	TRV	30[¶]	42
TRV	97	RRV-E	59	94

*Figures in this column are percentage protoplasts with RRV anti-gen aggregates.
[†]Protoplasts in this sample showed only diffuse cytoplasmic staining.
[¶]Most other protoplasts in this sample showed diffuse cytoplasmic staining.

brightly staining granules in doubly infected protoplasts, in contrast to the more usual faint generalized cytoplasmic staining in protoplasts infected only with RRV. In the early stages of infection fluorescent antibody to particles of TRV-CAM stained numerous granules in the cytoplasm of infected protoplasts, irrespective of whether they contained one or both viruses. In some tests with protoplasts at a later stage of infection with TRV-CAM the position of the chloroplasts was indicated by unstained spots outlined by stained cytoplasm. Thus in doubly infected protoplasts the distribution of RRV particle antigen came to resemble that of TRV-CAM particle antigen at an early stage of infection. These granules are thought to represent the aggregates of RRV and TRV-CAM particles revealed in the cytoplasm by electron microscopy of ultra-thin sections.

In the experiment cited in Table 1 many protoplasts made from *N. benthamiana* leaves, systemically infected with either RRV or TRV-CAM, were infected following inoculation with the other virus of the pair, resulting in the production of RRV antigen aggregates. Thus superinfection is not only possible but can result in an alteration in the distribution of RRV-E particle antigen in protoplasts which are already at a late stage of infection with this virus.

REFERENCES

Barker H. & Harrison B.D. (1977a) Infection of tobacco mesophyll protoplasts with raspberry ringspot virus alone and together with tobacco rattle virus. *Journal of General Virology* 35, 125-33.

Barker H. & Harrison B.D. (1977b) The interaction between raspberry ringspot and tobacco rattle viruses in doubly infected protoplasts. *Journal of General Virology* 35, 135-48.

Barker H. & Harrison B.D. (1978) Double infection, interference and superinfection in protoplasts exposed to two strains of raspberry ringspot virus. *Journal of General Virology* 40, 647-58.

Dawson J.R.O., Motoyoshi F., Watts J.W. & Bancroft J.B. (1975) Production of RNA and coat protein of a wild-type isolate and a temperature-sensitive mutant of cowpea chlorotic mottle virus in cowpea leaves and tobacco protoplasts. *Journal of General Virology* 29, 99-107.

Harrison B.D., Hutcheson A.M. & Barker H. (1977) Association between the particles of raspberry ringspot and tobacco rattle viruses in doubly infected *Nicotiana benthamiana* cells and protoplasts. *Journal of General Virology* 36, 535-9.

Kubo S., Harrison B.D., Robinson D.J. & Mayo M.A. (1975) Tobacco rattle virus in tobacco mesophyll protoplasts: infection and virus multiplication. *Journal of General Virology* 27, 293-304.

Otsuki Y. & Takebe I. (1969) Fluorescent antibody staining of tobacco mosaic virus antigen in tobacco mesophyll protoplasts. *Virology* 38, 497-9.

Otsuki Y. & Takebe I. (1976a) Double infection of isolated tobacco mesophyll protoplasts by unrelated plant viruses. *Journal of General Virology* 30, 309-16.

Otsuki Y. & Takebe I. (1976b) Double infection of isolated tobacco
    leaf protoplasts by two strains of tobacco mosaic virus.
    Biochemistry and Cytology of Plant-Parasite Interaction (Ed. by
    K. Tomiyama, J M. Daly, I. Uritani, H. Oku & S. Ouchi), pp. 213-22.
    Kodansha, Tokyo.
Takebe I. & Otsuki Y. (1969) Infection of tobacco mesophyll proto-
    plasts by tobacco mosaic virus. *Proceedings of the National Academy
    of Sciences, USA* 64, 843-8.
Takebe I., Otsuki Y. & Aoki S. (1968) Isolation of tobacco mesophyll
    cells in intact and active state. *Plant Cell Physiology* 9, 115-24.
Watts J.W., Dawson J.R.O. & King J.M. (1975) Mixed infection of
    tobacco protoplasts with cowpea chlorotic mottle virus and brome
    mosaic virus. *John Innes Institute Annual Report for 1975,*
    pp. 110-1.

Production of virus-free plants by tissue culture

D.G.A. WALKEY

National Vegetable Research Station,
Wellesbourne, Warwick CV35 9EF

INTRODUCTION

Since Morel & Martin (1952) first used meristem-tip culture to produce
virus-free dahlias, the technique has become widely adopted in horticulture. Unlike fungal and bacterial diseases, which may be eradicated
from crops by chemical spray treatments, the only practical methods of
eradicating viruses from vegetatively propagated species have been
tissue culture or thermotherapy, or a combination of both. When tissue
culture is used, in contrast to its use for rapid propagation, it is
only necessary for one healthy plantlet to be produced from the
diseased parent. This can then be propagated by conventional means or
by rapid tissue culture multiplication methods, as the situation demands.

   Various types of tissue may be cultured from the diseased parent,
including protoplasts, callus, nucellus, ovules, primordial floral
meristems and meristem-tips (Walkey, 1978). Of these, meristem-tips
have been found to be the most effective explants for the production
of virus-free plants of a wide range of economically important crops
(Quak, 1977; Walkey, 1978). Cultured on a suitable medium, meristem-
tips may be regenerated into plantlets more quickly than tissues from
other sources and, of even greater importance, the regenerated plants
usually retain the genetic characteristics of the parent. This is
probably due to the more uniformly diploid nature of the meristematic
cells (Murashige, 1974).

CLONE SELECTION AND EXPLANT CULTURE

A scheme for virus-free plant production is given in Fig. 1. At the
onset of any virus eradication programme it is important to select a
suitable parent clone for treatment. Although infected with the same
virus, individual clones or even individual plants may vary in their
vigour and capacity to multiply. This variation may be so pronounced
that clones freed of virus can be less vigorous than infected clones
of the same cultivar; consequently it is advisable to select for
treatment only clones that are known to propagate freely and to give

Ingram D.S. & Helgeson J.P. (1980) *Tissue Culture Methods for Plant Pathologists*

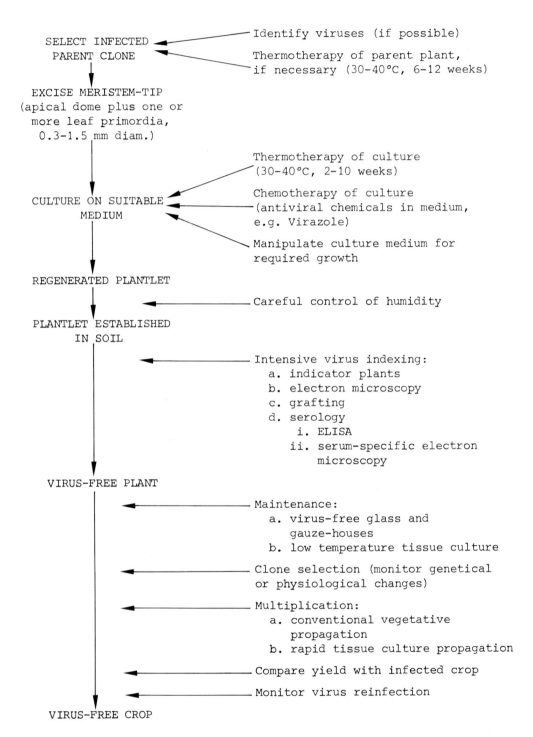

SELECT INFECTED
PARENT CLONE
— Identify viruses (if possible)

Thermotherapy of parent plant,
if necessary (30-40°C, 6-12 weeks)

EXCISE MERISTEM-TIP
(apical dome plus one or
more leaf primordia,
0.3-1.5 mm diam.)

CULTURE ON SUITABLE
MEDIUM

Thermotherapy of culture
(30-40°C, 2-10 weeks)

Chemotherapy of culture
(antiviral chemicals in medium,
e.g. Virazole)

Manipulate culture medium for
required growth

REGENERATED PLANTLET

Careful control of humidity

PLANTLET ESTABLISHED
IN SOIL

Intensive virus indexing:
a. indicator plants
b. electron microscopy
c. grafting
d. serology
     i. ELISA
     ii. serum-specific electron
         microscopy

VIRUS-FREE PLANT

Maintenance:
a. virus-free glass and
   gauze-houses
b. low temperature tissue culture

Clone selection (monitor genetical
or physiological changes)

Multiplication:
a. conventional vegetative
   propagation
b. rapid tissue culture propagation

Compare yield with infected crop

Monitor virus reinfection

VIRUS-FREE CROP

Figure 1.   Scheme for virus-free plant production by meristem-tip
culture.

relatively high yields.

Having selected a suitable clone, it is helpful to know which virus or viruses are present and to what extent the meristem-tip is infected. This information may be obtained by the same indexing methods as are used for virus testing the regenerated plantlet (Fig. 1), and although not essential, is of considerable assistance in eventually determining the disease status of the regenerated plantlet.

In general, the techniques used for excising and culturing the meristem-tip for virus-free plant production are similar to those used for tips in any rapid propagation programme. The essential factor is the selection of a suitable culture medium for vigorous, rapid growth; for most species, media based on that of Murashige & Skoog (1962) have been adequate. Details of explant removal, culture medium ingredients and conditions necessary for plantlet regeneration from meristem-tips have been extensively reviewed by Murashige (1974) and de Fossard (1976).

FACTORS CONTROLLING VIRUS ERADICATION

Meristem-tip size

When the meristem-tip culture technique was first used for virus-free plant production, it was assumed by many workers that all the regener-ated plants would be healthy, because virus was thought not to invade the meristematic cells of the bud. It was later found that in many cases this assumption was incorrect, and present evidence shows that viruses may invade meristematic cells to a varying extent, depending on the type of virus and the host species involved (Mori, 1977). For example, tobacco mosaic virus and potato virus X were reported to invade the meristem-tip to a greater extent than cucumber mosaic virus or potato virus Y. Success in obtaining virus-free plants by meristem-tip culture may therefore depend upon the initial size of tip removed for culture, as reported by Stone (1968) in carnations infected with carnation mottle, vein mottle and latent viruses. Tips varying in size from 0.1 to 2 mm in diameter have been cultured into plantlets. Most workers, however, have used tips between 0.5 and 1 mm in diameter. Such tips usually consist of the meristematic dome of cells plus two or more leaf primordia. Generally, the number of virus-free plantlets produced is inversely proportional to the size of the tip cultured. Thus, in some instances, it is possible to excise a meristem-tip free of the virus present in the infected parent and regenerate it into a healthy plant (Table 1).

In other cases viruses may be eradicated from meristem-tips during tissue culture, even when the tips can be shown to be infected at the time of excision, as in the elimination of carnation mottle virus from carnation (Hollings & Stone, 1964) and cherry leaf roll virus from tobacco (Walkey *et al.*, 1969). Mellor & Stace-Smith (1977) suggested that such *in vivo* eradication is caused by the metabolic disruption resulting from cell injury during the excision process; the smaller

the tip excised, the greater the injury and resultant disruption. It
seems probable that such *in vivo* virus eradication is more likely to
occur if small rather than large amounts of virus are present in the
tip.

In many plants it is impossible to excise a tip which is small
enough to avoid virus or for subsequent *in vivo* eradication, but which
is capable of regeneration to form a plantlet. In such instances
larger tips must be taken for culture, and these frequently contain
high concentrations of virus (Walkey & Webb, 1970). It is, however,
still possible to obtain virus-free plants from such infected material
by combining meristem-tip culture with high temperature treatments
(Table 1) or chemotherapy.

Table 1. Species which have been freed of virus by tissue culture.

Host	Viruses eradicated (heat treatment if applied)*
Allium sativum (garlic)	mosaic
Asparagus officinalis (asparagus)	unspecified
Armoracia lapathifolia (horse radish)	cauliflower mosaic (CaMV) turnip mosaic (TuMV)
Ananas sativus (pineapple)	unspecified
Brassica oleracea (cauliflower)	CaMV, TuMV
Caladium hortulanum (aroid)	dasheen mosaic
Chrysanthemum morifolium	virus B, vein mottle, greenflower, aspermy, stunt (35-38°C, 4-37 weeks)
Colocasia esculenta (taro)	dasheen mosaic
Cymbidium spp.	mosaic
Dahlia spp.	mosaic
Daphne odora	daphne virus S
Dianthus barbatus	ringspot, mottle, latent, vein-mottle
Dianthus caryophyllus (carnation)	ringspot, mottle, latent, streak, vein-mottle, etched-ring (35-40°C, 3-15 weeks)
Forsythia x intermedia	unspecified
Fragaria chiloensis (strawberry)	latent A & C, crinkle, yellow edge, vein chlorosis, vein banding (33-40°C, 4-7 weeks)

Table 1 (continued)

Host	Viruses eradicated (heat treatment if applied)*
Freesia spp.	mosaic, bean yellow mosaic
Geranium spp.	tomato ringspot
Gladiolus spp.	unspecified
Hippeastrum spp. (amaryllis)	mosaic
Humulus lupulus (hop)	mosaic, latent, prunus necrotic ringspot (40°C, 10-28 days)
Hyacinthus orientalis (hyacinth)	mosaic, lily symptomless
Ipomoea batatus (sweet potato)	internal cork, mosaic, mottle (27-28°C)
Iris spp.	latent, mosaic
Lavendula spp. (lavender)	dieback
Lilium spp.	cucumber mosaic (CMV), hyacinth mosaic, lily symptomless
Lolium multiflorum	rye-grass mosaic
Manihot utilissima (cassava)	mosaic, leaf distortion (35°C, 4-5 weeks)
Musa sapientum (banana)	CMV, unspecified (35-43°C, 14-15 weeks)
Narcissus tazetta (daffodil)	arabis mosaic (AMV), degeneration
Nasturtium officinale (watercress)	CMV, CaMV, TuMV
Nerine spp.	latent, unspecified
Nicotiana rustica (tobacco)	AMV, cherry leaf roll (CLRV), tobacco ringspot, alfalfa mosaic (32-40°C)
Nicotiana tabacum (tobacco)	tobacco mosaic (TMV)
Pelargonium spp.	unspecified
Petunia hybrida	TMV, tobacco necrosis
Ranunculus asiaticus	tobacco rattle, CMV
Rheum rhaponticum (rhubarb)	TuMV, CMV, CLRV, AMV, strawberry latent ringspot
Ribes grossularia (gooseberry)	vein banding (35°C, 2 weeks)
Rubus ideaus (raspberry)	mosaic
Saccharum officinarum (sugar cane)	mosaic
Solanum tuberosum (potato)	potato viruses A, G, M, S, X, Y, paracrinkle, spindle tuber (33-38°C, 4-18 weeks)
Xanthomosa brasiliense (cocoyam)	unspecified

* See reviews by Quak (1977) and Walkey (1978) for references

Thermotherapy

High temperature treatments have been widely used in virus-free plant
production (Nyland & Goheen, 1969). The treatment is usually carried
out on part of or on the whole of the infected parent plant by growing
it in a controlled temperature cabinet at 30-40°C for periods of 6-12
weeks. Normally no attempt is made to eradicate virus from the whole
plant, as healthy new shoots resulting from the high temperature treatment provide sufficient material for subsequent meristem-tip culture or
bud grafting (e.g. apples infected with chlorotic leafspot, stem
pitting or decline).

As an alternative to thermotheraphy of the infected parent plant it
is possible to eradicate viruses such as cucumber mosaic virus (CMV)
and alfalfa mosaic virus by culturing infected meristematic tissue at
30-40°C (Walkey, 1976). In both types of treatment it appears that
high temperature blocks virus replication and, since virus degradation
also takes place (Kassanis, 1957), the end result is complete eradication. Studies with CMV and alfalfa mosaic virus have suggested that at
least 45 days at 32°C or 9 days at 40°C are required for virus
eradication (Walkey, 1976), but the treatments required to eradicate
other viruses may vary.

The duration of the high temperature treatment is often critical.
If the treated tissues are returned to a lower temperature before the
virus is completely inactivated the remaining particles may multiply
to produce concentrations even higher than in the original infected
tissues (Walkey, 1976). The reason for this resurgence appears to be
the high temperature inactivation of a virus resistance factor in the
host.

Other temperature inactivation studies have shown that daily cycles
of high and low temperatures are frequently preferable to continuous
high temperature treatment, as they are less damaging to the tissues
(Walkey & Freeman, 1977). It has also been shown that virus degradation following blocking of virus synthesis may occur at low temperatures of about 5°C (Moskovets *et al.*, 1973).

Chemotherapy

It has been suggested that the presence of growth-promoting chemicals
in the culture medium will cause the eradication of virus during
meristem-tip culture (Quak, 1961), but there is little or no experimental evidence to support this. Recent experiments, in which
meristematic tissues infected with CMV were cultured on media
containing a range of cytokinin and auxin concentrations, showed that
these substances sometimes reduced virus concentration but did not
eradicate the virus (Cohen & Walkey, unpublished).

The latest chemotherapeutic studies suggest that the incorporation
of anti-metabolite chemicals such as Virazole (Ribavirin) in the
culture medium is likely to be more effective. These chemicals, in

common with high temperature treatments, block virus replication in infected tissues; presumably, while virus synthesis is stopped, degradation of existing virus continues until eradication occurs. Shepard (1977) reported that Virazole eradicated potato virus X in cultured tobacco shoots, and more recently it has been found that Virazole at concentrations of 50 and 100 mg/l will eradicate CMV from meristematic *Nicotiana rustica* L. tissues (Simpkins & Walkey, unpublished).

INDEXING, STOCK MAINTENANCE AND YIELD

Once the regenerated plantlet has been established in soil, it is essential to ensure that it is virus-free. Because many viruses have a delayed resurgence period, virus indexing must be carried out several times during the first year following culture, before the plant can be confidently used as nuclear stock material. In addition, regular tests should be carried out on the nucleus of virus-free plants retained for stock purposes.

Methods of virus testing depend on the species concerned and the facilities available. In the past grafting, electron microscopic examination of leaf and sap material, sap transmission to susceptible hosts and various serological tests have been used. Recently, however, the introduction of the ELISA (Enzyme Linked Immunosorbent Assay; Clark & Adams, 1977) and of serum-specific electron microscopy (Derrick, 1973) have provided methods of greater sensitivity.

In the past regenerated plantlets and virus-tested mother plants have generally been maintained in insect-proof glass or gauze houses. Recent studies, however, have shown that healthy stock material may be more readily stored for long periods by aseptic tissue culture at low temperatures (Mullin & Schlegel, 1976), a technique that may prove less expensive and time consuming.

Although most evidence shows that plants derived from meristem-tips usually exhibit little or no genetic differences from their virus-infected parents, it is essential to monitor small changes in the horticultural characteristics of regenerated plants. For example, past observations indicate that minor physiological changes may occur as a result of the absence of the virus; in apples minor variations in fruit colour and in the times of flowering and fruiting have been found (Campbell, unpublished), and in rhubarb major changes in the low temperature requirement to break dormancy have been reported (Case, 1973). Consequently, selection of the most horticulturally suitable clone following plantlet regeneration should be an important consideration.

Once a suitable clone has been selected and the plant multiplied to commercial quantities it is essential to conduct field trials to compare the yield and quality of the healthy and virus-infected crops. In addition, it is important that trials are carried out to monitor

rates of virus reinfection in various localities. The rate of
reinfection will be affected by the epidemiology of the virus concerned
and the degree of isolation the healthy crop can be given. In general,
it may be expected that aphid-borne viruses will be the first to
reinfect, particularly if the healthy crop is planted near reservoirs
of the virus. Information gained in such trials will help in determining the benefits likely to result from the use of virus-free
planting material.

REFERENCES

Case M.W. (1973) Rhubarb. *15th Annual Report of the Stockbridge
Experimental Husbandry Station, Yorkshire.*

Clark M.F. & Adams A.N. (1977) Characteristics of the microplate
method of enzyme-linked immunosorbent assay for the detection of
plant viruses. *Journal of General Virology* 34, 475-83.

de Fossard R.A. (1976) *Tissue Culture for Plant Propagators.*
University of New England Printing.

Derrick K.S. (1973) Quantitative assay for plant viruses using sero-
logically specific electron microscopy. *Virology* 56, 652-3.

Hollings M. & Stone O.M. (1964) Investigations of carnation viruses
I. Carnation mottle. *Annals of Applied Biology* 53, 103-18.

Kassanis B. (1957) The use of tissue culture to produce virus-free
clones from infected potato varieties. *Annals of Applied Biology*
45, 422-7.

Mellor F.C. & Stace-Smith R. (1977) Virus-free potatoes by tissue
culture. *Plant Cell, Tissue and Organ Culture* (Ed. by J. Reinert
& Y.P.S. Bajaj) pp. 616-35. Springer-Verlag, Berlin.

Morel G.M. & Martin C. (1952) Guérison de dahlias atteints d'une
maladie à virus. *Comptes rendus hebdomadaire des séances de
l'Académie des sciences, Paris* 235, 1324-5.

Mori K. (1977) Localisation of viruses in apical meristems and
production of virus-free plants by means of meristems and tissue
culture. *Acta Horticulturae* 78, 389-96.

Moskovets S.N., Gorbarenko N.I. & Zhuk I.P. (1973) The use of the
method of the culture of apical meristems in the combination with
low temperature for the sanitation of potato against mosaic virus.
Sel' skokhozjajstvennaja Biologija 8, 271-5.

Mullin R.H. & Schlegel D.E. (1976) Cold storage maintenance of strawberry meristem plants. *Horticultural Science* 11, 1004.

Murashige T. (1974) Plant propagation through tissue culture. *Annual
Review of Plant Physiology* 25, 135-66.

Murashige T. & Skoog F. (1962) A revised medium for rapid growth and
bioassays with tobacco tissue culture. *Physiologia Plantarum* 15,
473-97.

Nyland G. & Goheen A.C. (1969) Heat therapy of virus diseases of
perennial plants. *Annual Review of Plant Pathology* 7, 331-54.

Quak F. (1961) Heat treatment and substances inhibiting virus
multiplication in meristem culture to obtain virus-free plants.
Advances in Horticultural Science 1, 144-8.

Quak F. (1977) Meristem culture and virus-free plants. *Plant Cell, Tissue and Organ Culture* (Ed. by J. Reinert & Y.P.S. Bajaj), pp. 598-615. Springer-Verlag, Berlin.

Shepard J.F. (1977) Regeneration of plants from protoplasts of potato virus X infected tobacco leaves. II. Influence of Virazole in the frequency of infection. *Virology* 78, 261-6.

Stone O.M. (1968) The elimination of four viruses from carnation and sweet william by meristem-tip culture. *Annals of Applied Biology* 62, 119-22.

Walkey D.G.A. (1976) High temperature inactivation of cucumber and alfalfa mosaic viruses in *Nicotiana rustica* cultures. *Annals of Applied Biology* 84, 183-92.

Walkey D.G.A. (1978) *In vitro* methods for virus elimination. *Frontiers of Plant Tissue Culture, 1978* (Ed. by T.A. Thorpe), pp. 245-54. International Association for Plant Tissue Culture, Calgary.

Walkey D.G.A., Fitzpatrick J. & Woolfitt J.M.G. (1969) The inactivation of virus in cultured tips of *Nicotiana rustica* L. *Journal of General Virology* 5, 237-41.

Walkey D.G.A. & Freeman G.H. (1977) Inactivation of cucumber mosaic virus in cultured tissues of *Nicotiana rustica* L. by diurnal alternating periods of high and low temperature. *Annals of Applied Biology* 87, 375-82.

Walkey D.G.A. & Webb M.J.W. (1970) Tubular inclusion bodies in plants infected with viruses of the NEPO type. *Journal of General Virology* 7, 159-66.

Pathogen elimination and *in vitro* plant storage in forage grasses and legumes

P.J. DALE, V.A. CHEYNE & S.J. DALTON
Welsh Plant Breeding Station,
Plas Gogerddan, Aberystwyth SY23 3EB

INTRODUCTION

Perennial forage grass and legume cultivars are usually based on 8-30 selected genotypes from which seeds are multiplied over about four sexual generations until there is a sufficient quantity for release commercially. It is important to keep the mother plants alive and in a healthy condition for many years so that they can be used for seed production when required. Conventionally the mother plants are main-tained in a glasshouse or field where they are continually exposed to pests and diseases and are often difficult to keep alive, especially the naturally short lived species.

Shoot-tip culture of grasses and legumes is being evaluated for the elimination of important pathogens and for establishing plant storage cultures. This report describes methods of shoot-tip culture for pathogen elimination and plant storage under conditions of minimal growth for several agriculturally important forage grasses and legumes. Most experience has been with forage grasses, and the mother plants of several cultivars bred at the Welsh Plant Breeding Station are already being stored *in vitro* (Jones & Dale, 1978). The information presented for legume pathogen elimination and storage is preliminary. No problems of genetic or cytological instability have been encountered to date.

MATERIALS AND METHODS

Forage grasses

Preparation of plant material   Separate vegetative plants into individual tillers, remove any dead leaves and trim the roots to 0.5 cm and the shoots to 5 cm. Wash the tillers in running tap water to remove soil. Surface sterilize the tillers in a glass or plastic dish with a lid by briefly pre-rinsing (30 s) in absolute ethanol and then submerg-ing in a sodium hypochlorite solution (5-7% w/v available chlorine) for 15 min. Add 1 drop of Tween 80 per 100 ml of the sodium hypochlorite solution as a wetting agent and shake the dish several times during

Ingram D.S. & Helgeson J.P. (1980) *Tissue Culture Methods for Plant Pathologists*

Table 1. The proportion of plants regenerated from shoot tips and tiller buds in which virus could not be detected.

Host	Virus	Explant	Shoot tips and tiller bud length (mm)				Maximum explant size* (mm)
			0-0.5	0.6-1.0	1.1-2.0	2.1-5.0	
Lolium multiflorum	ryegrass† mosaic	shoot tip	40/45 (89%)	17/20 (85%)	11/22 (50%)	0/10 (0%)	0.3
		tiller bud	27/39 (69%)	11/33 (33%)	7/13 (54%)	1/13 (8%)	0.2
	barley¶ yellow dwarf	shoot tip	1/1 (100%)	4/5 (80%)	4/9 (44%)	3/9 (33%)	0.8
		tiller bud	-	2/3 (67%)	6/10 (60%)	4/5 (80%)	-
Dactylis glomerata	cocksfoot** streak	shoot tip	3/3 (100%)	3/3 (100%)	-	4/4 (100%)	4.8
		tiller bud	4/4 (100%)	4/4 (100%)	1/1 (100%)	0/7 (0%)	1.1
	cocksfoot** mild mosaic	shoot tip	3/3 (100%)	3/3 (100%)	-	4/4 (100%)	4.8
		tiller bud	4/4 (100%)	4/4 (100%)	1/1 (100%)	7/7 (100%)	5.0
	cocksfoot** mottle	shoot tip	1/1 (100%)	1/1 (100%)	-	-	0.7
		tiller bud	3/3 (100%)	3/3 (100%)	-	1/2 (50%)	2.2

*All shoot tips or tiller buds up to this size have so far given plants free of the virus.
†The results of a more extensive study since Dale (1977b), where all the plants regenerated from shoot tips up to 1.1 mm long were free of ryegrass mosaic virus.
¶Dale & Dalton, unpublished data.
**Dale, 1979.

surface sterilization. Working in a laminar flow cabinet, rinse the
tillers six times in sterile water and drain well.

Tillers are best prepared on the day of culturing but can be stored
aseptically for several days in a refrigerator (*c.* 4°C) without obvious
deterioration.

Preparation of culture medium   Prepare Murashige & Skoog's (1962)
basal medium with the addition of 30 g/l sucrose and 0.2 mg/l kinetin.
Adjust the pH to 5.6, add 5 g/l agar (Sigma Chemical Co., London; type
IV) and autoclave at 121°C for 15 min. Pour 10 ml aliquots of medium
into 85 x 28 mm sterile glass universal tubes or universal plastic tubes
(Sterilin Ltd, Teddington, Middlesex). This medium has been selected
for *Lolium multiflorum* (Dale, 1980), but is also being successfully used
for other forage grass species.

Dissection and culturing   Using a stereo microscope (magnification
x10) in a laminar flow cabinet, dissect the shoot tips with flamed
forceps and needle. Excise tips up to 0.5 mm long and place them on
the nutrient medium. Incubate the cultures at 25°C with continuous
white fluorescent light at 6000 lx. Plantlets regenerate in these
conditions over 1-2 months. The yield of plantlets from shoot tips
depends on the species (Dale, 1977a), and in most cases is 30-90%.

Elimination of pathogens   Pests, fungi and bacteria are usually
eliminated if aseptic techniques are observed when the shoot tips are
cultured. If not, the nutrient medium in most cases becomes colonized
by contaminant micro-organisms, and the culture must be discarded.

It is important that the mother plants stored *in vitro* are healthy,
thereby increasing their chances of survival in culture and enabling
them to grow vigorously when moved to soil for seed production. The
efficiency of eliminating virus pathogens by shoot-tip culture, as in
many other species, depends on the size of tip cultured. The effect of
culturing shoot tips of different sizes is presented in Table 1, where
data on the use of tiller buds as explants have been included for
comparison. Tiller buds lie in the axils of the leaves and are exposed
by stripping down successive leaves to the base of the stem.

For a given explant size, culturing tiller buds is generally less
effective for virus elimination than culturing shoot tips, presumably
because tiller buds carry more leaf material. It is obviously desirable
to establish cultures from small shoot tips and, where possible, these
should not be greater than 0.5 mm long.

Plant storage   When the regenerated plantlets have shoots 2-4 cm long
place them under low temperature storage conditions (Fig. 1). After 10-
11 months at reduced temperature (1 year from initial culturing) provide
fresh nutrients by subculturing to new culture medium. Use the same
medium for initial plantlet regeneration, storage and subculture. As
there is no requirement for virus elimination at subculturing, use any
explant capable of giving a new plantlet. Tiller buds are easy to

dissect and handle but shoot tips, tiller bases and even nodes can also be used (Dale, 1980). To obtain tiller bases remove the plantlet from its culture tube, separate into individual tillers, and trim the roots to 1-2 mm and the shoot to 2-5 mm from the base of the stem. Internode extension occurs in some plants in culture and nodes can be excised. Nodes produce plantlets from the apical shoot tip (the highest node) or from tiller buds.

Generally 10 plantlets of each genotype are stored *in vitro*, each in a separate culture tube. When plantlets are taken for seed multiplication or are lost through contamination the stocks can be replenished at subculturing, since about 7-10 explants are available from each stored plantlet.

To transfer plantlets from storage to soil, first move the culture tubes from cold storage to the regenerative conditions (25°C, continuous light) for a few days. Then remove the plantlets from their culture tubes, rinse the roots in running tap water to remove agar and transplant them into a sterilized, soil-based potting compost. Trim the leaves to 5 cm and cover individual plants with transparent glass or plastic beakers or polyethylene film for one week. Water thoroughly after transplanting and keep the plants at 15-20°C in a glasshouse until they become well established (2-3 weeks). Over 95% of plantlets can be successfully transferred to soil in this way.

The following species have been stored *in vitro* at the Welsh Plant Breeding Station for 3 or more years: *L. multiflorum*, *Lolium perenne*, *L. multiflorum* x *L. perenne*, *Dactylis glomerata*, *Festuca pratensis*, *Festuca arundinacea*, *Phleum pratense*.

Forage legumes

The following methods are being developed for white clover, red clover (diploid and tetraploid) and lucerne.

Preparation of plant material   Remove stems from the plant, leaving sufficient behind to allow the plant to recover its growth in soil. Wash the stems thoroughly in running tap water. Remove the leaves with their petioles and surface sterilize the stems by submerging them in a sodium hypochlorite solution (5-7% w/v available chlorine) for 15 min using a glass or plastic dish with a lid. Add 1 drop of Tween 80 per 100 ml of the sodium hypochlorite solution and agitate the dish several times during sterilization. Working in a laminar flow cabinet rinse the stems six times in sterile water and drain well.

Preparation of culture media   Several basal culture media and growth regulator combinations have been tested for regenerating plantlets from shoot tips in forage legumes (Cheyne, 1979). Of those tested the best are Blaydes' (1966) and Gamborg's B$_5$ (Gamborg *et al.*, 1968).

Prepare Blaydes' medium with 0.2 mg/l indole 3-ylacetic acid and 0.2 mg/l 6(γ,γ-dimethylallylamino) purine riboside for the clovers and

Gamborg's B$_5$ basal medium with 0.2 mg/l 1-naphthaleneacetic acid for lucerne.   In each case add 30 g/l of sucrose, adjust the pH to 5.5, add 5 g/l Sigma (type IV) agar and autoclave for 15 min at 121°C.   Dispense 10 ml aliquots into 30 ml capacity Sterilin universal plastic tubes.

Dissection and culturing   Using a stereo microscope (magnification x10) in a laminar flow cabinet, expose the apical and lateral shoot tips with a sterile scalpel and fine forceps.   Excise the shoot tips at the required length and place them onto the culture medium.   Regenerate plantlets by incubating the cultures at 25°C under continuous white fluorescent light at 6000 lx.   Using this protocol with shoot tips 0.3-4.0 mm long, 62-82% of clover shoot tips and 80% of lucerne shoot tips have regenerated plants.

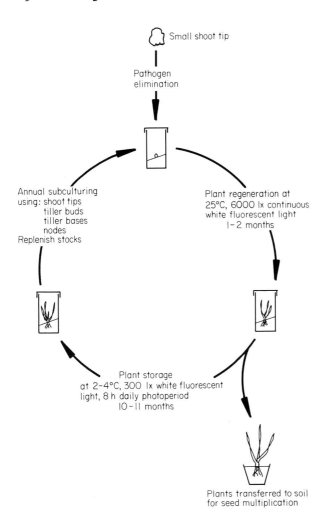

Figure 1. The *in vitro* plant storage system.

Elimination of pathogens and plant storage   Various pathogens are
important in legumes. Phyllody mycoplasma is a particular problem in
breeding and research as it transforms inflorescences into leafy
structures, thus preventing seed production. Clover red leaf mycoplasma
causes a decline in plant vigour and may lead to plant death. Preliminary evidence indicates that both pathogens can be eliminated by
culturing shoot tips up to 1 mm long.

The storage system described for grasses is being tested for legumes.
To date, white clover and lucerne have been stored at low temperature
for 1 year and red clover for 6 months. Plantlet survival under these
conditions has been high. All the legume species can be transferred
effectively from culture to soil (with *c.* 90% success) using the method
described for grasses.

REFERENCES

Blaydes D.F. (1966) Interaction of kinetin and various inhibitors in
    the growth of soybean tissue. *Physiologia Plantarum* 19, 748-53.
Cheyne V.A. (1979) Tissue culture studies in legumes. *M.Sc. Thesis,
    University of Wales, Aberystwyth.*
Dale P.J. (1977a) Meristem tip culture in *Lolium, Festuca, Phleum* and
    Dactylis. Plant Science Letters 9, 333-8.
Dale P.J. (1977b) The elimination of ryegrass mosaic virus from *Lolium
    multiflorum* by meristem tip culture. *Annals of Applied Biology* 85,
    93-6.
Dale P.J. (1979) The elimination of cocksfoot streak virus, cocksfoot
    mild mosaic virus and cocksfoot mottle virus from *Dactylis glomerata*
    by shoot tip and tiller bud culture. *Annals of Applied Biology* 9,
    285-8.
Dale P.J. (1980) Development of *in vitro* storage for *Lolium multiflorum.
    Annals of Botany* 45, 497-502.
Gamborg O.L., Miller R.A. & Ojima K. (1968) Nutrient requirements of
    suspension cultures of soybean root cells. *Experimental Cell
    Research* 50, 151-8.
Jones M.Ll. & Dale P.J. (1978) Plant maintenance in culture. *Report of
    the Welsh Plant Breeding Station for 1977*, pp. 46-7.
Murashige T. & Skoog F. (1962) A revised medium for rapid growth and
    bioassays with tobacco tissue cultures. *Physiologia Plantarum* 15,
    473-97.

Culture of *Pelargonium* hybrids from meristems and explants: chimeral and beneficially-infected varieties

A.C. CASSELLS*, G. MINAS† & R. LONG†

*Department of Botany, University College, Cork
†Department of Horticulture, Wye College
(University of London), Wye, Kent TN25 5AH

INTRODUCTION

Meristem tip culture, sometimes preceded by heat treatment of the donor plant, has been used successfully to eliminate *Xanthomonas pelargonii* and several viral pathogens from *Pelargonium* hybrids, including the commercially important "geraniums" (Hamdorf, 1976). The application of tissue culture research to the commercial propagation of "geraniums" in the USA and Europe has stimulated the development of tissue culture cloning methods. The goal has been to develop a tissue culture bank, easily maintained and at low risk of reinfection, which could be rapidly proliferated to provide new élite mother stock each season. This is a development of the original strategy, whereby single plants produced by tip culture were maintained in insect-free environments until they had been disease indexed, so that clean individuals could be vegetatively propagated to produce élite mother stock. This strategy is both expensive and demanding, as the highest standards of hygiene are required throughout.

Meristem culture has been preferred to explant and callus culture as a cloning procedure for *Pelargonium* as these have in some cases led to genetic variation in the progeny plants (Skirvin & Janick, 1976). However, meristem culture (Hamdorf, 1976) may lead to disease elimination, which poses a problem when beneficially-infected varieties are to be propagated. The use of explant cultures can circumvent this problem (Cassells, 1979). We report here tissue culture media and methods for the cloning of *Pelargonium* from meristems and from petiolar explants. The examples discussed are of chimeral and beneficially-infected varieties. The potential role of callus culture in the production of new genotypes is considered.

MATERIALS AND METHODS

Plant material

Use the following *Pelargonium* cultivars: Crocodile (syn. Sussex lace);

Ingram D.S. & Helgeson J.P. (1980) *Tissue Culture Methods for Plant Pathologists*

the ivy-leaf "geranium" (*P. x peltatum*) characterized by its clear
veins, which have been attributed to virus infection or infection
with a virus-like agent; and Mme Salleron, a variegated leaf zonale
"geranium" (*P. x zonale*). The inheritance of variegation has been
investigated by Tilney-Bassett (1974). Grow plants in a glasshouse
(min. temp. 15°C) in Irish peat moss supplemented regularly with Bio
P Base fertilizer (Pan Britannica Industries, Hertfordshire).

Meristem cloning

Remove apical buds from actively growing plants in spring and early
summer (autumn and winter isolations are unsuccessful) and, without
surface sterilization, remove the outer leaves and leaf primordia under
a dissecting microscope in a laminar air-flow cabinet. Culture the

Table 1. Tissue culture media.

Medium	Constituents	mg/l	g/l
Meristem medium (K4)	MS* basal medium without growth substances		4.71
	ammonium nitrate		0.825
	sodium dihydrogen phosphate		0.15
	casein hydrolysate		1.0
	sucrose		30.0
	indoleacetic acid	2.0	
	gibberellic acid (GA$_3$)	1.0	
	kinetin	4.0	
	adenine sulphate	50.0	
	meso-inositol	100.0	
	agar		6.0
Differentiation medium (Explant medium; Cassells, 1979)	MS* basal medium without growth substances		4.71
	sucrose		30.0
	zeatin	1.0	
	agar		6.0
Rooting medium	MS* basal medium without growth substances		2.36
	sucrose		15.0
	kinetin	0.01	
	indoleacetic acid	0.1	
	agar		6.0

pH adjusted to 5.8
*Murashige & Skoog (1962)

apical domes with the first pairs of leaf primordia in 60 ml glass bottles containing 10 ml of K4 medium (see Table 1) and incubate in a growth room at 22°C with a 16 h photoperiod of 5W/m^2. If a halo of pigment develops around the explants move them to a new position on the medium. Establishment and first proliferation will occur within 2 weeks, and after *c.* 6 weeks the proliferating cultures should be subdivided and subcultured onto fresh K4 medium. Repeat this procedure until the desired number of adventitious meristems has been produced.

Adventitious shoot development occurs when the cultures are allowed to develop without subdivision. Excise individual shoots at random and place on 10 ml of rooting medium (Table 1) in 60 ml jars. Roots will develop in 2-3 weeks, and the plantlets should then be transferred to potting compost in an insect-proof propagator at 18°C for a further 2 weeks. Finally transfer them to the glasshouse in insect-proof cages as appropriate, and grow on to flowering or the 20 leaf stage.

In our experiments meristem cultures of Mme Salleron did not differ qualitatively from those of 30 other non-variegated varieties studied on the defined media described. Meristem cultural responses of ivy-leaved varieties, including Crocodile, were similar to those of zonale varieties. The need to maintain the plants of the varieties described in this paper in insect-proof environments for several months prevented extensive cloning. However, comparative studies of mass cloning of the following varieties were carried out: Irene, Brook's Purple and Paul Crampel (zonale varieties); and Mme Layal (a regal *Pelargonium; P.* x *domesticum*). A proliferation of 10 000 adventitious stems in 6 months was achieved, starting with six meristems and involving three subdivisions. Batches of 200 of each of these varieties were grown on to flowering, and both leaf and flower characteristics studied. No variation was found in Mme Layal but 15% of the Irene, Brook's Purple and Paul Crampel showed leaf distortions.

Explant culture

Harvest petioles in spring and surface sterilize in 80% v/v ethanol for 30 s and in 10% v/v commercial sodium hypochlorite solution (Domestos; Lever Bros, Liverpool) for 15 min. Wash the explants in sterile distilled water, trim back to 2 cm and place in 60 ml jars containing 10 ml of differentiation medium (Table 1). Culture the explants as above. Adventitious shoot formation will occur within *c.* 6 weeks. Individual adventitious shoots should then be separated, rooted, established in soil and maintained as above.

In our experiments explant cultures of Mme Salleron developed sparse callus and produced a mean of two adventitious shoots per explant. The ivy-leaf explants produced shoots direct, with less callus formation than above. In both examples the suppressive effect of the first formed shoots on further adventitious shoot formation was observed (Cassells, 1979).

COMPARISON OF METHODS

To compare the meristem and explant cloning procedures 21 plantlets
derived from meristem culture and 30 derived from explant culture of
Crocodile were grown to flowering. All plants derived from explant
culture showed clear veins, although in some symptoms were delayed for
up to 1 month from the time of establishment in soil. Clear veins were
absent from the meristem culture plants up to the time of flowering.
This result is illustrated in Fig. 1.

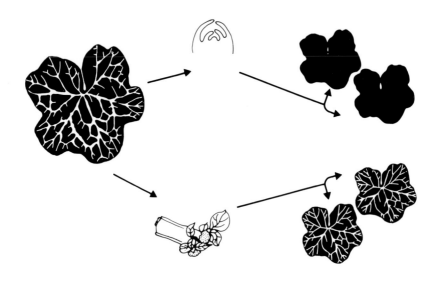

Figure 1. Meristem culture (top) and explant culture (bottom) of
the beneficially-infected "geranium" cv. Crocodile. Vein-clearing
is transmitted in explant culture but not in meristem culture.

In another experiment 75 adventitious shoots from meristem cultures
of Mme Salleron were selected for rooting; of these 18 have so far been
established and grown into plants. All retained the typical leaf
variegation. Explant cultures produced only albino and all-green
adventitious shoots. The green shoots were successfully rooted and 50
plants were grown on to the 20 leaf stage. No leaf variegation was
detected in any. This is illustrated in Figure 2.

CONCLUSIONS

The sample results presented above illustrate fundamental differences
between meristem culture and explant (and callus) culture in relation
to the propagation of chimeral and beneficially-infected varieties:
there may be a high frequency of transmission of pathogens through
explant culture and elimination in meristem culture; and chimeral

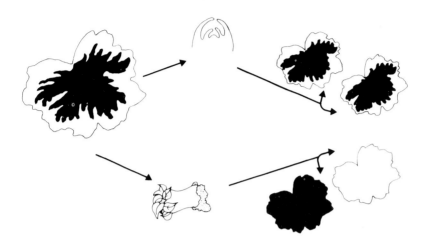

Figure 2. Meristem culture (top) and explant culture (bottom) of
the variegated "geranium" cv. Mme Salleron. The chimera is
perpetuated in meristem culture and broken down in explant culture.

breakdown may occur in explant culture while chimerals are perpetuated
in meristem culture.

Assuming the validity of these generalizations, it follows that
symptomless viral infections may also be transmitted with high frequency
via explant culture (Cassells & Long, this volume) to produce clonally
infected stock; this is a particular risk in crops like *Pelargonium*,
in which many virus symptoms are expressed only at certain times in
the season. Explant culture may also lead to considerable genetic
variation in the progeny plants, particularly when the explant is
derived from a variety that is traditionally propagated vegetatively.
This variation may result from the breakdown of visible chimeras of
the sort described here or from the segregation of mutations which
have accumulated in the somatic cells of the plant (D'Amato, 1977).
The latter situation has been reported by Skirvin & Janick (1976) to
occur in scented-leaf geranium. These workers showed that genetic
variation was high in the first generation progeny derived from
explants but that second generation progeny showed less variation.
It has been suggested that plants derived from vegetative embryos
produced in explant/callus culture may be derived from single cells,
and thus the progeny plants are initially genetically uniform (Cassells,
1979). The significance for the plant pathologist in the separation
of somatic cell lines in explant/callus culture is that amongst these
there may be new disease-resistant lines; thus this procedure may be
valuable in the search for new sources of disease resistance.

REFERENCES

Cassells A.C. (1979) The effect of 2,3,5-triiodobenzoic acid on
    caulogenesis in callus cultures of tomato and *Pelargonium*.
    Physiologia Plantarum 46, 159-64.
D'Amato F. (1977) Cytogenetics of differentiation in tissue and cell
    cultures. *Plant Cell, Tissue, and Organ Culture* (Ed. by J. Reinert
    & Y.P.S. Bajaj), pp. 343-56. Springer Verlag, Berlin.
Hamdorf G. (1976) Propagation of *Pelargonium* varieties by stem-tip
    culture. *Acta Horticulturae* 59, 143-51.
Murashige T. & Skoog F. (1962) A revised medium for rapid growth
    and bioassay with tobacco tissue cultures. *Physiologia Plantarum*
    15, 473-97.
Skirvin R.M. & Janick J. (1976) Tissue culture-induced variation in
    scented *Pelargonium* spp. *Journal of the American Society for
    Horticultural Science* 101, 281-90.
Tilney-Bassett R.A.E. (1974) The control of plastid inheritance in
    Pelargonium III. *Heredity* 33, 353-60.

Tobacco explant culture as a screening system for antiviral chemicals

A.C. CASSELLS* & R. LONG†

*Department of Botany, University College, Cork
†Department of Horticulture, Wye College
(University of London), Wye, Kent TN25 5AH

INTRODUCTION

The strategy for mass cloning of plants by micropropagation depends on disease-indexing the donor plant, and eliminating any diseases found, before setting up the explant or meristem cultures. Details of tissue culture media have been published for the cloning of most commercially important species (Reinert & Bajaj, 1977). Fungal, bacterial and mycoplasmal pathogens can be successfully eliminated by the addition of the appropriate antibiotics to the culture medium, but the cloning potential of micropropagation will be realized only when virus elimination can be easily achieved.

At present, virus exclusion from the meristem and/or differential thermal stability of virus and host tissue are used to produce individual virus-free "elite" stock plants for subsequent cloning (Matthews, 1970). However, these procedures are not universally efficacious, some viruses being particularly invasive or having relatively high thermal stability (Hollings, 1965; Nyland & Goheen, 1969). Furthermore, low virus titre or slow systemic movement may necessitate repeated virus testing of the progeny plant before one can be reasonably confident of virus elimination; during this time there is risk of reinfection.

Studies on antiviral chemicals *in vitro* using plant tissues have frequently yielded promising results. However, when the same chemicals have been tested in the field, results have been equivocal at best. The rationale of the logical synthesis of antiviral chemicals has centred mainly on the synthesis of competitive inhibitors of viral enzymes or host processes involved in replication. Effective *in vitro*, where inhibitory concentrations can be controlled and toxicity/inhibitory balance achieved, inhibition is lost *in vivo* where the inhibitor concentration changes with redistribution and growth.

In tobacco explant culture, regeneration can occur either via adventitious shoot or vegetative embryo formation. Plantlets derived from explants taken from donor plants infected with potato virus Y

Ingram D.S. & Helgeson J.P. (1980) *Tissue Culture Methods for Plant Pathologists*

(PVY) or cucumber mosaic virus (CMV) are all infected. Morphogenesis, from single cells or small groups of cells, and the high frequency of virus transmission, make explant culture a novel system for screening potential antiviral chemicals. This chapter describes a method for the application of tobacco explant culture to the screening of antiviral chemicals using Virazole (1-β-D-ribofuranosyl-1,2,4-triazole-3-carboxamide) as an example.

MATERIALS AND METHODS

Growth of plants

Grow *Nicotiana tabacum* L. cv. xanthi-nc plants in a glasshouse in Irish Peat Moss supplemented with Bio P Base (Pan Britannica Industries Ltd, Hertfordshire) fertilizer. In Wye the mean incoming solar radiation, corrected for absorption by the glass, is 1.3 MJ/m^2/day (range 0.5 to 1.6 MJ/m^2/day). Extend the day length to 15 h or longer as appropriate with supplementary radiation from mercury vapour lamps (HPLR, 400 W; Philips Ltd, London). Take explants from plants at *c.* the 15 leaf stage.

Explant culture

Surface sterilize petiole explants in 80% v/v aqueous ethanol for 1 min and then in 10% v/v aqueous commercial sodium hypochlorite solution (Domestos; Lever Bros, Liverpool) for 15 min. Wash in sterile water, cut into 1 cm lengths and place on the culture medium.

The basic medium used should be the mineral salts, vitamins and sucrose solution of Murashige & Skoog (1962) supplemented with 9.4 µM zeatin (mixed isomers) and 0.8% w/v agar and adjusted to pH 5.8. Autoclave the full medium at 121°C for 15 min. Culture explants in a growthroom (22°C, 5W/m^2, 16 h photoperiod). Embryo-like structures will form in the areas of sparse callus development and groups of bud-like structures will arise direct from epidermal cells and wing tissue. Later these will develop as shoots.

Screening for antiviral activity

Grow seedlings of *N. tabacum* cv. xanthi-nc as described above and mechanically inoculate at the 3-4 leaf stage with either PVY or CMV (wild strains). Excise petioles from the systemically infected plants when they reach the 12-14 leaf stage (*c.* 35 cm in height) and surface sterilize and culture as above. Add sterile filtered Virazole to the medium after autoclaving (autoclaving Virazole reduces its activity). Take shoots at random from these cultures when they reach *c.* 1 cm in length. The time of sampling varies with the concentration of Virazole used in the medium, from 6 weeks at 20.5 µM Virazole to *c.* 22 weeks at 205 µM Virazole.

Excise shoots from the explants and place these direct into

unsterilized potting compost on a propagating bench in insect-proof boxes. After rooting (*c.* 2-3 weeks later) pot on the plantlets and grow in insect-proof cages. When the plants reach the 4 leaf stage screen them for virus. Test for CMV on *Chenopodium amaranticolor* by local lesion assay and for PVY by leaf dip electron microscopy (Noordam, 1973). Plants which give a negative reaction should be grown on to the 15 leaf stage and retested (Fig. 1).

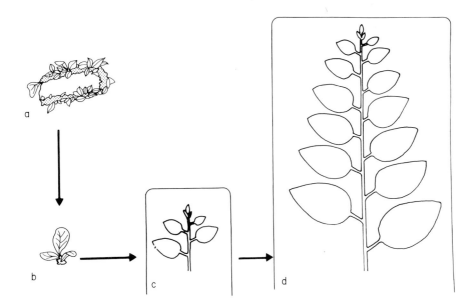

Figure 1. Tobacco petiole explant culture as a screening system for antiviral chemicals: a, induction of adventitious shoot formation; b, separation of adventitious shoots and transfer to insect-proof cages for rooting; c, virus testing of progeny plants at the 4 leaf stage (following removal from the tissue cultures the plants are maintained in insect-proof cages); d, if the virus test at the 4 leaf stage is negative the plants are grown on to the 15 leaf stage and retested.

Additional comments on procedures

Length of photoperiod   Experiments have shown that petiole explants taken from donor plants given a 24 h photoperiod for 7 days prior to harvesting the explants produce significantly more adventitious shoots than petiole explants from plants given a 16 h photoperiod only.

Time of harvest   Experiments using plants grown with a 16 h photoperiod have shown that petioles harvested at 5 a.m. (end of the dark period) produce significantly less adventitious shoots per explant than petioles harvested at 12 noon and 9 p.m. (end of the light period).

Petiole position   Petioles taken from positions near the top of plants produce more adventitious shoots than petioles taken from the middle or bottom of plants.

Typical results (see Table 1)

Explants from both PVY and CMV infected donor plants cultured in the absence of Virazole gave rise to adventitious shoots, all of which were infected. This result has been obtained repeatedly. PVY-infected explants, cultured in the presence of 20.5, 41 and 205 μM Virazole, produced adventitious shoots which were 99%, 87.5% and 0% infected respectively, while CMV-infected explants cultured on the same concentrations of Virazole produced adventitious shoots which were 79%, 74% and 0% infected respectively. When infected explants, after bud formation on medium containing 205 μM Virazole, were transferred to Virazole-free medium prior to excising the shoots, the percentage of infected shoots increased from 0% to 44% and 33% for PVY and CMV respectively. In interpreting the data it should be noted that virus transmission from infected to non-infected adventitious shoots is possible in the cultures before and during excision and rooting.

Table 1. The percentage of infected adventitious shoots produced in explant cultures in the presence of different concentrations of Virazole. Data are for the mean of replicate experiments (12 replicates per treatment). The controls were in all cases 100% infected.

Virus	Virazole concentration (μM)	Number of plants screened	% virus infected
PVY	20.5	52	99
	41.0	48	87.5
	205.0	19	0
	205.0	43*	44
CMV	20.5	69	79
	41.0	54	74
	205.0	16	0
	205.0	77*	33*

PVY, potato virus Y
CMV, cucumber mosaic virus
*Partial replicate experiment in which the explants were cultured on Virazole containing medium bud formation, then transferred to Virazole medium before subsequent excision of the shoots.

CONCLUSIONS

In this work, and in work on the regeneration of plants from protoplasts isolated from potato virus X-infected plants (Shepard, 1977), it has been shown that Virazole in the differentiation medium may result in the production of virus-free adventitious shoots. The effect of Virazole has been shown to be reversible, and the data suggest an effect on virus movement into the adventitious buds and/or vegetative embryos from which the adventitious shoots are derived. Previous studies have suggested that Virazole may inhibit a host function affecting virus maturation (Sarver & Stoller, 1978). If Virazole acts as proposed here, it may have broad spectrum activity and commercial application in the micropropagation of virus-free stock plants.

REFERENCES

Cassells A.C. (1979) The effect of 2,3,5-triiodobenzoic acid on caulogenesis in callus cultures of tomato and *Pelargonium*. *Physiologia Plantarum* 46, 159-64.
Hollings M. (1965) Disease control through virus-free stock. *Annual Review of Phytopathology* 3, 367-96.
Matthews R.E.F. (1970) *Plant Virology*. Academic Press, New York.
Murashige T. & Skoog R. (1962) A revised medium for rapid growth and bioassay with tobacco tissue cultures. *Physiologia Plantarum* 15, 473-97.
Noordam D. (1973) *Identification of Plant Viruses*. PUDOC, The Hague, Netherlands.
Nyland G. & Goheen A.C. (1969) Heat therapy of virus diseases of perennial plants. *Annual Review of Phytopathology* 7, 331-54.
Reinert J. & Bajaj Y.P.S. (1977) *Plant Cell, Tissue and Organ Culture*. Springer Verlag, Berlin.
Sarver N. & Stollar V. (1978) Virazole prevents production of sindbis virus and virus-induced cytopathic effect in *Aedes albopictus* cells. *Virology* 91, 267-82.
Shepard J.F. (1977) Regeneration of plants from protoplasts of potato virus X-infected tobacco leaves. II. Influence of Virazole on frequency of infection. *Virology* 78, 261-6.

Section 4
Methods for use with fungi, bacteria and nematodes

The establishment of dual cultures of downy mildew fungi and their hosts

D.S. INGRAM

Botany School, Downing Street, Cambridge CB2 3EA

INTRODUCTION

Dual cultures of obligate biotrophs and callus tissues of their hosts
are potentially of value for the maintenance of supplies of aseptic
inoculum, the cloning of isolates, the safe international transport of
isolates, studies of host-parasite interaction and attempts to produce
axenic cultures (Ingram, 1976, 1977).

The first dual cultures of a downy mildew fungus and its host were
established by Morel (1944), who inoculated callus tissue of vine
(*Vitis vinifera*) with aseptic zoospores of *Plasmopara viticola*. Since
then methods have been devised for growing a number of other host-downy
mildew combinations (Table 1). This chapter gives details of a method
that has proved successful with the temperate combination *Beta vulgaris*
(sugar beet)-*Peronospora farinosa* f. sp. *betae,* and then deals with
problems pertinent to the establishment of dual cultures of other
combinations.

METHOD (SUGAR BEET AND *P. FARINOSA*)

Establishment of host callus tissue cultures and parasite inoculum

Before initiating dual cultures it is necessary to establish callus
tissue cultures of sugar beet and to prepare inoculum of the parasite.
This stage may take several weeks. The methods involved are given
below.

Host   Establish vigorously growing callus tissue cultures of sugar
beet according to the methods outlined by Helgeson (this volume).
Ingram & Joachim (1971) obtained good growth of callus tissues of a
range of sugar beet genotypes on two media: the CM medium of Ingram
(1969a), which included the growth substances 2,4-dichlorophenoxyacetic
acid (2,4-D; 6.0 mg/l), 1-naphthaleneacetic acid (NAA; 0.1 mg/l) and
coconut milk (150 ml/l); and the basal medium of Murashige & Skoog
(1962) supplemented with agar (7.0 g/l), kinetin (1.5 mg/l) and either
2,4-D (6.0 mg/l), or NAA (1.0 mg/l) or 4-chloro-2-oxobenzothiazolin-3-

ylacetic acid (benazolin; 5.0 mg/l). *P. farinosa* grew more vigorously in association with callus tissues maintained on the medium containing the complex extract than on the defined media.

Incubate non-infected sugar beet callus cultures at 25°C without light. (Note that dual cultures require a temperature of 15°C with low intensity white light; see below.)

Ideal culture vessels for growing calluses to be used in dual culture work are plugged 100 ml Erlenmeyer flasks, or Petri dishes sealed with Parafilm.

Parasite   Detach pairs of cotyledons, together with *c.* 1 cm of hypocotyl, from 7- to 14-day old seedlings of a susceptible sugar beet cultivar and lay these out on crinkled filter paper moistened with distilled water in clear, lidded plastic boxes (125 x 80 x 20 m). Make a suspension of conidia of *P. farinosa* from infected sugar beet plants, wash twice with distilled water by low speed centrifugation to remove the water soluble inhibitor of germination and resuspend in distilled water (*c.* 3 x 10^5/ml). Inoculate the cotyledons by spraying with the suspension of sporangia until thoroughly wetted and incubate at 15°C in low intensity white light (*c.* 1000 lx; 16 h photoperiod). Sporulation usually occurs after 7-14 days of incubation. Use the resulting conidia to inoculate a fresh batch of cotyledons and then maintain the fungus by regular transfer.

Initiation of dual cultures

Method 1   Surface sterilize infected cotyledon pairs 72 h after inoculation by dipping them rapidly in 95% v/v ethanol and then immersing them for 5 min in a solution of sodium hypochlorite (1% available chlorine). Wash the cotyledons three times in sterile distilled water and then place individual pairs aseptically on Murashige & Skoog basal medium (i.e. without growth substances) in Petri dishes sealed with Parafilm. Incubate at 15°C in low intensity white light (*c.* 1000 lx; 16 h photoperiod). Sporulation will occur in 7-14 days.

At sporulation select uncontaminated cotyledons (usually 5-50%) and transfer each pair aseptically to the surface of a sugar beet callus culture already established in a plugged 100 ml Erlenmeyer flask or in a Petri dish sealed with Parafilm. Incubate at 15°C in low intensity white light (*c.* 1000 lx; 16 h photoperiod). In most cases the fungus will spread from the cotyledons and become established in the callus tissue within about 10 days. Eventually a weft of aerial mycelium will form over the entire surface of each infected culture.

Method 2   Prepare uncontaminated sporulating cotyledons as for Method 1. Excise 0.5 cm segments of infected tissue, place these direct onto one of the callus tissue culture media (i.e. with growth substances) in appropriate culture vessels and incubate at 15°C in low intensity white light (*c.* 1000 lx; 16 h photoperiod). Callus will form at the cut surfaces and in many cases will contain systemic mycelium of *P. farinosa*.

Maintenance of dual cultures

The balance in culture between *P. farinosa* and its host is usually
such that dual cultures grow and enlarge as rapidly as non-infected
calluses. Maintain dual cultures, therefore, by regular (e.g. two
monthly) transfer of explants of infected tissue to fresh medium, where
they will continue to proliferate. Alternatively, encourage sporulation on a dual culture by adding a small quantity of sterile distilled
water to the culture vessel, to increase the humidity; then, after
incubating the culture for 24-28 h, transfer the resulting conidia in
a droplet of water to a fresh callus culture.

Confirmation of pathogenicity of the fungus grown in dual culture

At regular intervals confirm the pathogenicity of *P. farinosa* grown in
dual culture by inoculating detached cotyledons of the host with a
droplet of distilled water containing either a fragment of infected
callus tissue or conidia taken from a dual culture, and incubate at 15°C
in low intensity white light (*c.* 1000 lx; 16 h photoperiod). Providing
that the fungus has retained its pathogenicity infections will be
established and sporulation will occur on the cotyledons in 7-10 days.

OTHER DOWNY MILDEWS

All host-downy mildew combinations must be regarded as being unique as
far as dual culture is concerned, and it is necessary to devise a
separate protocol for each one. The references listed in Table 1 will
provide a guide to the techniques that have been used so far. The
following comments may also be helpful.

Host tissue cultures

The chapters by Helgeson and by Thomas *et al.* in this volume provide
basic information.

Contaminant-free inoculum

Methods similar to that described above for *P. farinosa* have provided
contaminant-free spores of *Peronospora parasitica* (Ingram, 1969b) and
Bremia lactucae (Mason, 1973). Yields of uncontaminated cotyledons
were 30-50% for *P. parasitica-Brassica* spp. and 60-95% for *B. lactucae-*
lettuce. In the case of *P. viticola* Morel (1944, 1948) surface
sterilized mature infected vine leaves and allowed the fungus to grow
out through the leaf surface again. Then, having removed spores
aseptically in sterile distilled water, he allowed zoospores to form
and used these to inoculate vine callus cultures. A reasonable number
of uncontaminated cultures were obtained in this way. Griffin & Coley-
Smith (1968) obtained contaminant-free sporangia of *Pseudoperonospora
humuli* from surface sterilized segments of systemically infected hop
stems. The incidence of bacterial contamination in these segments was
usually less than 50%. Tiwari & Arya (1969) infected callus tissues of

Table 1. Downy mildew fungi grown in association with host tissue cultures

Fungus	Host	References
Bremia lactucae Regel*	*Lactuca sativa* L.* (lettuce)	Mason, 1973
Peronospora farinosa (Fr.) Fr. f.sp. *betae*	*Beta vulgaris* L. (sugar beet)	Ingram & Joachim, 1971
Peronospora parasitica (Pers. ex Fr.) Fr.	*Brassica* spp. (cabbage, turnip etc.)	Nakamura, 1965; Ingram, 1969b
Peronospora tabacina Adams	*Nicotiana tabacum* L. (tobacco)	Izard *et al.*, 1964
Plasmopara viticola (Berk. & Curt.) Bebl. & de Toni	*Vitis vinifera* L. (vine)	Morel, 1944, 1948
Pseudoperonospora humuli (Miy. & Tak.) Wilson	*Humulus lupulus* L. (hop)	Griffin & Coley-Smith, 1968
Sclerophthora macrospora (Sacc.) Thirum. Shaw & Naras	*Eleusine coracana* (L.) Gaertn. (finger millet)	Safeeulla, 1976
Sclerospora graminicola (Sacc.) Schroet.	*Pennisetum typhoideum* (Burm.) Stapf et C.E. Hubb. (pearl millet)	Tiwari & Arya, 1967, 1969; Arya & Tiwari, 1969; Safeeulla, 1976; Bhat *et al.*, 1980
Sclerospora sacchari T. Miyake (≡*Peronosclerospora sacchari* (T. Miyake) C.G. Shaw)	*Saccharum officinarum* L. (sugar cane)	Chen *et al.*, 1979
Sclerospora sorghi Weston & Uppal (≡*Peronosclerospora sorghi* (Weston & Uppal) C.G. Shaw)	*Sorghum vulgare* Pers.	Safeeulla, 1976

*The tissues became infected, but the fungus did not proliferate (see text).

pearl millet with *Sclerospora graminicola* by placing on them small fragments of systemically infected flower tissue aseptically dissected from the apical region of a mature infected plant.

In some instances it has been possible to obtain dual cultures of downy mildew fungi and their hosts direct by placing explants of systemically infected tissue on the surface of a culture medium and then allowing the two organisms to proliferate together. This was the case with sugar cane and *Sclerospora sacchari* (Chen *et al.*, 1979).

Balance in culture

Downy mildew fungi differ in their ability to live in balance with host tissue cultures. Thus, for example, whereas dual cultures of sugar beet and *P. farinosa* live in complete harmony (see above) callus tissues of *Brassica* spp. are killed within 7-14 days of infection with *P. parasitica* (Ingram, 1969b). In the latter case dual cultures can only be maintained by inoculating non-infected host callus cultures with explants of tissue from moribund dual cultures at regular intervals.

In some instances the reverse situation may obtain, with a vigorous host tissue culture outgrowing the fungus. This was sometimes the case with dual cultures of sugar beet and *P. farinosa* grown on the defined culture medium (Ingram & Joachim, 1971).

The balance between a downy mildew fungus and its host in culture is likely to be influenced by the degree of biotrophy inherent in the fungus and by the relative vigour of the two organisms. The latter factor may be susceptible to modification by variation in the physical components of the culture environment (e.g. temperature) and in the composition of the culture medium (e.g. levels of growth substances).

Abnormal resistance

In some cases it may be difficult or impossible to establish dual cultures because the cultured tissues of the host are abnormally resistant to the fungus. Thus, for example, despite growing callus tissues of a range of cultivars of lettuce on a range of media, it has so far been impossible to establish dual cultures with *B. lactucae*. The accumulation of a fungitoxic compound in the lettuce tissues in response to infection is apparently responsible (Mason, 1973; Ingram, unpublished data). It should also be noted that the balance of growth substances in the culture medium may have a marked effect on the expression of resistance to fungi in cultured tissues (Helgeson & Haberlach, this volume).

REFERENCES

Arya H.C. & Tiwari M.M. (1969) Growth of *Sclerospora graminicola* on callus tissues of *Pennisetum typhoides* and in culture. *Indian Phytopathology* <u>22</u>, 446-52.

Bhat S.S., Safeeulla K.M. & Shaw C.G. (1980) Growth of *Sclerospora graminicola* in host tissue cultures. *Transactions of the British Mycological Society* (in press).

Chen W.-H., Liu M.-Ch. & Chao Ch.-Y. (1979) The growth of sugarcane downy mildew fungus in tissue culture. *Canadian Journal of Botany* 57, 528-33.

Griffin M.J. & Coley-Smith J.R. (1968) The establishment of hop tissue cultures and their infection by downy mildew *Pseudoperonospora humuli* (Miy. & Tak.) Wilson under aseptic conditions. *Journal of General Microbiology* 53, 231-6.

Ingram D.S. (1969a) Growth of *Plasmodiophora brassicae* in host callus. *Journal of General Microbiology* 55, 9-18.

Ingram D.S. (1969b) The susceptibility of *Brassica* callus to infection by *Peronospora parasitica*. *Journal of General Microbiology* 58, 391-401.

Ingram D.S. (1976) Growth of biotrophic parasites in tissue culture. *Encyclopedia of Plant Physiology*, New Series (Ed. by R. Heitefuss & P.H. Williams), Volume 4, pp. 743-59. Springer-Verlag, Berlin.

Ingram D.S. (1977) Applications in plant pathology. *Plant Tissue and Cell Culture* (Ed. by H.E. Street), pp. 463-500. Blackwell Scientific Publications, Oxford.

Ingram D.S. & Joachim I. (1971) The growth of *Peronospora farinosa* f. sp. *betae* and sugar beet callus tissues in dual culture. *Journal of General Microbiology* 69, 211-20.

Izard C., Lacharpagne J. & Schiltz P. (1964) Behaviour of *Peronospora tabacina* in tissue cultures, and role of the leaf epidermis. *Annales de la Direction des Études et de l'Equipment*, Section 2, pp. 95-9. Service d'Exploitation Industrielle des Tabacs et des Allumettes.

Mason P.A. (1973) Studies on the biology of *Bremia lactucae* Regel. *Ph.D. Thesis, University of Cambridge*.

Morel G. (1944) Le développement du mildiou sur des tissus de vigne cultures *in vitro*. *Comptes rendus hebdomadaire des séances de l'Académie des sciences, Paris* 218, 50-2.

Morel G. (1948) Recherches sur la culture associée de parasites obligatoires et de tissus végétaux. *Annales des Éphiphyties (série Pathologie Végétale)* 14, 1-112.

Murashige T. & Skoog F. (1962) A revised medium for rapid growth and bioassays with tobacco tissue cultures. *Physiologia Plantarum* 15, 473-97.

Nakamura H. (1965) The use of tissue cultures in the study of obligate parasites. *Proceedings of an International Conference on Plant Tissue Culture*, pp. 535-9. McCutchan Publishing Co., Berkeley, California.

Safeeulla K.M. (1976) *Biology and Control of the Downy Mildews of Pearl Millet, Sorghum and Finger Millet*. Downy Mildew Research Laboratory, Mysore, India.

Tiwari M.M. & Arya H.C. (1967) Growth of normal and diseased *Pennisetum typhoides* tissues infected with *Sclerospora graminicola* in tissue culture. *Indian Phytopathology* 20, 356-68.

Tiwari M.M. & Arya H.C. (1969) *Sclerospora graminicola* axenic culture. *Science* 163, 291-3.

Culture of *Plasmodiophora brassicae* in host callus tissue

S.T. BUCZACKI

National Vegetable Research Station,
Wellesbourne, Warwick CV35 9EF

INTRODUCTION

Plasmodiophora brassicae Woron. is an obligately biotrophic, non-mycelial and zoosporic organism classified within the Plasmodiophorales, a grouping of uncertain affinities (Karling, 1969). It causes clubroot of crucifers, an economically important disease of temperate climate vegetables (Buczacki, 1979). Although the life cycle of the organism is still imperfectly known, the evidence seems increasingly to support the version proposed by Ingram & Tommerup (1972). The primary stage of the life cycle, in which zoosporangial plasmodia form, is known to occur in a wide range of host plants from several families. The secondary stage, however, in which cystogenous plasmodia proliferate and produce resting bodies in the cortical and associated root tissues, seems to be restricted to the family Cruciferae. This secondary stage is character-ized by a rapid and massive increase in the size and number of infected and neighbouring cells leading to pronounced and characteristic root gall development. It appears that the galling process is related, at least in part, to the formation in the tissue of abnormally large quantities of indole 3-ylacetic acid (IAA) as a result of the degrada-tion of naturally occurring indole glucosinolates such as glucobrassicin (Butcher *et al.*, 1974; Ockendon & Buczacki, 1979).

The first successful dual culture of *P. brassicae* in naturally infected root tissue of cabbage (*Brassica oleracea* L. var. *capitata* L.) was reported briefly by Strandberg *et al.* (1966) and more fully by Williams *et al.* (1969). The culture medium was slightly modified from that described by Linsmaier & Skoog (1965) after Murashige & Skoog (1962) and transfers of tissue onto fresh medium were usually made at 7- to 12-day intervals. By such frequent transfer of rapidly prolifer-ating calluses it was possible to maintain *P. brassicae* in a plasmodial state without any formation of resting cysts. Ingram (1969a, 1969b) confirmed these findings and further showed that *P. brassicae* could also be maintained for long periods in a plasmodial state in host callus on a medium containing coconut milk and 2,4-dichlorophenoxyacetic acid (2,4-D). Only when the latter ingredient was omitted was the organism likely to undergo cystogenesis. Dekhuijzen & Overeem (1971) confirmed earlier

Ingram D.S. & Helgeson J.P. (1980) *Tissue Culture Methods for Plant Pathologists*

observations that infected callus, unlike healthy tissue, was independent of an exogenous supply of cytokinins and hormones for its continued
growth, although new tissue was produced more prolifically when these
substances were present. Dekhuijzen & Overeem were uncertain of the
origin of the endogenous growth substances in infected tissues but drew
attention to the observation (Ingram, 1969b) that when the parasite had
been lost from a callus clone, that clone did not retain the ability to
grow independently of an external supply of such substances. Minor
modifications to the medium composition have been made by different
workers, mainly with respect to the relative amounts of the growth
factors, and it seems probable that different tissue lines, especially
those from different host species, have slightly differing requirements
for optimum growth. The media described below are those that have been
found, after prolonged testing at Wellesbourne, to give the most
satisfactory results with a limited range of tissues.

Although tissue culture techniques have been used in studies of the
growth of P. brassicae in host callus, they have been relatively little
used in investigations of clubroot disease. Tommerup & Ingram (1971)
compared the life cycle of P. brassicae in tissue culture with that in
intact roots. They showed that all known stages of the life cycle
occurred within the callus, including those that would normally be found
in the root hairs or root epidermal cells. They recorded the seemingly
aberrant germination of resting cysts in situ but subsequently showed
(Ingram & Tommerup, 1972) that this could also occur in normal galled
roots and could similarly lead to the formation of zoosporangial
plasmodia in deep-seated host cells. There appear therefore to be no
basic features of the life cycle of P. brassicae in callus culture that
are not capable of occurring in root galls.

The greatest single practical advantage to research on P. brassicae
conferred by tissue culture is that it enables pathogen inoculum to be
obtained free of contaminating organisms. Thus resting cysts may be
obtained in sterile suspension for inoculation of plants under aseptic
conditions, while callus also provides a sterile source of plasmodia
when these are required in isolation for physiological or other studies:
Keen et al. (1969) and Dekhuijzen (1975, 1976) extracted plasmodia
either mechanically or enzymically and measured their rates of respir-
ation and nutrient uptake. Plasmodia isolated from callus might
themselves prove amenable to culture in nutrient solution, a technique
that could produce valuable information about the nutritional requirements of P. brassicae. Buczacki & Moxham (1979) described some
tentative explorations in this direction.

Hitherto it has proved impossible to infect healthy brassica callus
artificially with inoculum either of resting cysts (Buczacki, unpublished data) or plasmodia (Dekhuijzen, 1975). Moreover, Dekhuijzen
(personal communication) failed to induce a transfer of P. brassicae
from infected into healthy callus when the two were juxtaposed. It is
of interest, however, that Sacristan & Hoffmann (1979) infected both
sterile stem embryo cultures of Brassica napus in a liquid medium
(Murashige & Skoog, 1962) without hormones, and sterile callus derived

from single cells in suspension in a similar medium supplemented with 2,4-D and kinetin. The inoculum was a suspension of well-washed resting spores.

METHODS AND MATERIALS

The major problems in the callus culture of *P. brassicae* are associated with its initiation from galled roots, when the avoidance of bacterial contamination is very difficult. For this reason the successful initiation of proliferating cultures from more than 10% of initials should not be expected. Many workers will evolve their own procedures for sterilizing gall tissue and their own experience will indicate the optimum size for each initial, but the following procedures are those that have been found most reliable at Wellesbourne.

Production of galled plants

Firm young galls from artificially inoculated plants must be used. Naturally infected roots are unlikely to give satisfactory initials, largely because the level of bacterial and other contamination will be too high to eliminate entirely. Inoculum of resting cysts should be prepared by the homogenization of root galls and the filtration and centrifugation of the resultant homogenate (Buczacki, 1973). Ten-day old seedlings should be inoculated by dipping their roots in cyst suspension before transplanting them into sterile sand fed with a standard nutrient solution. Alternatively, the cyst suspension may be incorporated into the moistened sand and the seeds sown into this medium. Apart from the use of sand rather than a soil-based medium, the subsequent culture of the inoculated plants in a glasshouse should be based on any of the procedures catalogued by Dixon (1976), who also gives details of appropriate inoculum levels.

Whilst successful cultures have been established from a number of cruciferous species, cabbage and turnip (*B. campestris* L. var. *rapifera* Metz.) have consistently given the best results. The optimum length of time after inoculation until suitable galls form will vary slightly with the conditions under which the plants are grown, but 42 days will be adequate in most circumstances.

Preparation of initials

The plants should be uprooted, then the galls cut off, gently brushed to remove adhering sand, trimmed free of fibrous roots and washed in running tap water for several hours. Only creamy white galls should be selected for use; any with brownish discoloured patches should be discarded. Cubes of tissue not larger than 1 cm^3 should be cut from the galls and shaken gently in sterile distilled water on a flask shaker for approximately 30 min. The shaking should be repeated twice, using fresh water each time, and the gall pieces then immersed for 1 min in 95% ethanol. They should then be transferred without washing to 1% mercuric chloride for 4 min and all subsequent operations should be performed on a sterile bench. Each tissue piece should be individually washed by hand-shaking for about 1 min in a covered sterile beaker containing sterile distilled water. This washing should be repeated 10 times for each tissue piece using a clean, sterile beaker each time. Thereafter each piece should be transferred to a few ml of sterile distilled water in an individual sterile Petri dish and trimmed with a scalpel to remove the external tissues. The remaining tissue should be cut into cubes of approximately 8 mm^3 and each cube placed on 30 ml of culture medium in a 100 ml conical flask stoppered with cotton wool and metal foil.

Newly established cultures should be monitored by microscope examination of hand sections, to ensure that they are infected.

Culture medium

In our experience the most suitable medium for the culture of callus infected with *P. brassicae* is that of Linsmaier & Skoog (1965) modified by replacing IAA with 1-naphthaleneacetic acid (NAA; 0.3 mg/l) and by adding kinetin (1 mg/l). The pH should be adjusted to 5.8 before autoclaving. Growth will occur in the absence of kinetin, and in some instances it may be found more satisfactory to omit this substance until the calluses are established.

Culture conditions

Flasks should be incubated in the dark at 22-24°C and actively proliferating callus transferred every 10-12 days to fresh medium if the organism is to be maintained in a plasmodial state. If the cultures are not transferred in this way, resting cyst production may be expected in 4-5 weeks.

REFERENCES

Buczacki S.T. (1973) Glasshouse evaluation of some systemic fungicides for control of clubroot of brassicae. *Annals of Applied Biology* 74, 85-90.
Buczacki S.T. (1979) *Plasmodiophora brassicae. CMI Descriptions of Pathogenic Fungi and Bacteria* No. 621.
Buczacki S.T. & Moxham S.E. (1979) Clubroot of brassicas. Studies of *Plasmodiophora brassicae* biology. *Report of the National Vegetable Research Station for 1978*, p. 75.
Butcher D.N., El-Tigani S. & Ingram D.S. (1974) The role of indole glucosinolates in the clubroot disease of the Cruciferae. *Physiological Plant Pathology* 4, 127-41.
Dekhuijzen H.M. (1975) The enzymatic isolation of secondary vegetative plasmodia of *Plasmodiophora brassicae* from callus tissue of *Brassica campestris. Physiological Plant Pathology* 6, 187-92.
Dekhuijzen H.M. (1976) The role of growth hormones in clubroot formation. *Mededelingen van de Faculteit Landbouwwetenschappen Rijksuniversiteit Gent* 41/2, 517-23.
Dekhuijzen H.M. & Overeem J.C. (1971) The role of cytokinins in

clubroot formation. *Physiological Plant Pathology* 1, 152-62.

Dixon G.R. (1976) Methods used in Western Europe and the USA for testing *Brassica* seedling resistance to clubroot (*Plasmodiophora brassicae*). *Plant Pathology* 25, 129-34.

Ingram D.S. (1969a) Growth of *Plasmodiophora brassicae* in host callus. *Journal of General Microbiology* 55, 9-18.

Ingram D.S. (1969b) Abnormal growth of tissues infected with *Plasmodiophora brassicae*. *Journal of General Microbiology* 56, 55-67.

Ingram D.S. & Tommerup I.C. (1972) The life history of *Plasmodiophora brassicae* Woron. *Proceedings of the Royal Society, B* 180, 103-12.

Karling J.S. (1969) *The Plasmodiophorales*, 2nd edition. Hafner, New York.

Keen N.T., Reddy M.N. & Williams P.H. (1969) Isolation and properties of *Plasmodiophora brassicae* plasmodia from infected crucifer tissues and from tissue culture callus. *Phytopathology* 59, 637-44.

Linsmaier E.M. & Skoog F. (1965) Organic growth factor requirements of tobacco tissue cultures. *Physiologia Plantarum* 18, 100-29.

Murashige T. & Skoog F. (1962) A revised medium for rapid growth and bioassays with tobacco tissue cultures. *Physiologia Plantarum* 14, 473-97.

Ockendon J.G. & Buczacki S.T. (1979) Indole glucosinolate incidence and clubroot susceptibility of three cruciferous weeds. *Transactions of the British Mycological Society* 72, 156-7.

Sacristan M.D. & Hoffman F. (1979) Direct infection of embryogenic tissue cultures of haploid *Brassica napus* with resting spores of *Plasmodiophora brassicae*. *Theoretical and Applied Genetics* 54, 129-32.

Strandberg J.O., Williams P.H. & Yukawa Y. (1966) Monoxenic culture of *Plasmodiophora brassicae* with cabbage tissue. *Phytopathology* 56, 903.

Tommerup I.C. & Ingram D.S. (1971) The life-cycle of *Plasmodiophora brassicae* Woron. in *Brassica* tissue cultures and in intact roots. *New Phytologist* 70, 327-32.

Williams P.H., Reddy M.N. & Strandberg J.O. (1969) Growth of noninfected and *Plasmodiophora brassicae* infected cabbage callus in culture. *Canadian Journal of Botany* 47, 1217-21.

The recalcitrant powdery mildews — attempts to infect cultured tissues

K.J. WEBB* & J.L. GAY

Department of Botany and Plant Technology,
Imperial College of Science and Technology,
London SW7 2BB

INTRODUCTION

An attempt was recently made to infect cells in culture with powdery
mildew fungi in order to obtain an experimental system more amenable to
manipulation than natural infections for the study of disease physiology.
Although the objective was not attained, valuable data were accumulated;
this report gives details of the methods used and summarizes the trials
made. It is intended as a guide for those who wish to repeat the work
or who wish to extend it to other powdery mildew fungi.

Although most powdery mildews only infect epidermal cells, several
lines of evidence indicate that they are not wholly specialized in
infecting one cell type. Two genera (*Phyllactinia* and *Leveillula*)
infect internal cells of leaves, and mesophyll infections by epidermal
mildews have also been reported. The latter have sometimes occurred
without any special treatment of the host (*Erysiphe pisi*; Gil, 1976) or
have developed after removal of the epidermis (*Erysiphe cichoracearum*;
Schnathorst, 1959), after mechanical or chemical injury (*Erysiphe
graminis*; Salmon, 1905, 1906) or following heating (*E. graminis*,
E. cichoracearum and *Sphaerotheca fuliginea*; Jarvis, 1964, 1968, and
personal communication). Attempts to infect tissue cultures of *Vitis*
spp., *Helianthus annuus* and *Rosa* spp. with their respective powdery
mildews *Uncinula necator* (Morel, 1948), *E. cichoracearum* (Heim & Gries,
1953) and *Sphaerotheca pannosa* (Mence & Hildebrandt, 1966) have also
been reported; the first two reports describe limited success.

Here, we give methods to test the abilities of tissue cultures of
Rosa spp. (rose), *Acer pseudoplatanus* (sycamore), *Triticum aestivum*
(wheat) and *Pisum sativum* (pea) to support their respective powdery
mildews *S. pannosa*, *Uncinula aceris*, *E. graminis* f.sp. *tritici* and
E. pisi. Complementary experiments with detached pea leaves are also
described.

---

*Present address: Department of Plant Biology, University of Birmingham,
PO Box 363, Birmingham B15 2TT.

Ingram D.S. & Helgeson J.P. (1980) *Tissue Culture Methods for Plant Pathologists*

MATERIALS AND METHODS

Maintenance of stocks of callus and suspension cultures

Grow stock cultures in 100 ml or 250 ml Erlenmeyer flasks containing 20 ml or 50 ml medium respectively and plugged with non-absorbent cotton wool. Cultures grown under a 16 h photoperiod at 25°C can be subcultured or transferred every 3 weeks (rose and sycamore) or 6 weeks (wheat and pea). Maintain callus cultures on agar medium by transferring small pieces of callus (*c.* 250 mg fresh weight), and suspension cultures by transferring cells plus medium with an automatic syringe (10 ml into 50 ml fresh medium). Seal the flasks with sterile aluminium foil squares. Maintain suspension cultures on a rotary shaker (125 rev/min).

Rose and sycamore   Long established cell lines of both species were used in our experiments. Callus cultures were obtained from Leicester University and suspension cultures were initiated from them. Cultures of Paul's Scarlet rose (PSR) were maintained on MX_2 medium (Nash & Davies, 1972) supplemented with 2,4-dichlorophenoxyacetic acid (2,4-D; 1.0 mg/l) and kinetin (0.5 mg/l). Cells of the AM line of sycamore (Bayliss & Gould, 1974) were isolated from cultures provided by D.T.A. Lamport (Lamport & Northcote, 1960) and were maintained on Heller's medium (Heller, 1953) with the modifications described in Stuart & Street (1969), and with the addition of 2,4-D (1.0 mg/l) and kinetin (0.25 mg/l). These suspension cultures comprised single cells and small aggregates of cells.

Wheat   Long established cell lines of the cultivar Maris Ranger originating from either seedling roots (MR4r) or cotyledons (MR4c) were obtained from J.F. O'Hara (O'Hara & Street, 1978). New cultures were also established from excised embryos of the susceptible varieties Morris Butler (MB3e) and Hobbit (H1e). Good growth was achieved on Murashige & Skoog (MS) medium (Murashige & Skoog, 1962) supplemented with 2,4-D (1.0 mg/l). These cultures exhibited a high degree of cellular organization, having numerous nodules containing vascular elements, and also retained the ability to produce roots on withdrawal of 2,4-D. Such cultures do not readily dissociate to produce cell suspensions in liquid medium.

Pea   New cultures were established from the susceptible cultivar Onward. Cultures were initiated from leaves (PL_1, PL_2 and PL_4) or petioles of young (YP_1) or older (OP_1, OP_2 and P_2) glasshouse-grown pea plants and also from entire seedlings (PSO_3). B_5 medium (Gamborg *et al.*, 1968) supplemented with 1-naphthaleneacetic acid (NAA; 2.0 mg/l) and kinetin (2.0 mg/l) supported the initial growth, the hormonal levels being reduced to 1.0 mg/l NAA and 0.5 mg/l kinetin on further transfer. These lines showed different morphological characteristics, e.g. friable or compact, but none exhibited organogenesis.

Maintenance of powdery mildew fungi

Use *S. pannosa* and *U. aceris* direct from rose and sycamore plants grown

in the field. To extend the natural seasonal availability of the fungus,
infected sycamore seedlings and rose bushes can be potted and transferred to a glasshouse at 25°C with a 16 h photoperiod.

To maintain the mildews *E. pisi* and *E. graminis* f.sp. *tritici* in the
glasshouse, shake old infected leaves or plants over fresh stocks of
healthy plants every 3 weeks. Purify cultures of these two mildews by
the successive transfer of a few conidia onto surface sterilized
detached leaves in culture as follows. Surface sterilize the leaves by
immersion in a 10% sodium hypochlorite solution (providing 1.2% free
chlorine) for 10 min, followed by 3 washes in sterile distilled water.
Place the leaves, adaxial surface uppermost, on water-agar supplemented
with 1.0 mg/l benzylaminopurine (BAP) in an unsealed Petri dish and
incubate at 20°C with a 16 h photoperiod. After they have dried (*c.* 24
h), inoculate the leaves with conidia (initially from plants, then from
leaf cultures) using a dry sterile mounted needle. After several such
transfers the conidia harbour relatively little contamination, as can be
determined by plating free conidia onto nutrient agar medium.

Twenty-four hours prior to experimental inoculation of tissue
cultures remove the old dead conidia by blowing a stream of air (sterile
in the instance of the purified cultures) over the colonies with a
Pasteur pipette. Use the newly formed viable conidia in all experiments.

Isolation of waxy substances and natural cuticle of pea

These procedures are modified from Yang & Ellingboe (1972).

Waxy substances   Extract wax by dipping pea leaves 3 times, 10 s each
time, into chloroform. Reconstruct a wax layer, with approximately the
same thickness as natural wax layers, on an agar surface with an area
equal to that of the extracted leaves. Allow the remaining chloroform
to evaporate completely and float the resultant wax layer on sterile
water. Wash well and use to overlay the callus by floating on a drop of
sterile water on the callus. Allow to evaporate leaving a film of wax
over the callus.

Natural cuticle   Remove the lower epidermis of pea leaves with fine
forceps. Place leaf and epidermal strips, cuticle side uppermost, on a
solution of 0.4% Macerozyme R10 and 1.2% cellulase R10 (Yakult Manufacturing Co., Japan). Leave for 18 h, or until the underlying cells are
loosened from the cuticle, and then wash the isolated cuticles with
sterile distilled water. The purity of the preparation can be ascertained by transmission electron microscopy of sectioned material.
Overlay on callus as described above.

EXPERIMENTAL

Use aseptic techniques throughout the following experiments. Inoculate
cultures at known times in the diurnal cycle using viable conidia either
on a dry sterile mounted needle or in a drop of filter-sterilized FC 43

Table 1. Summary of infection trials with wheat, pea, rose and sycamore cultures.

Experimental variables	Host, Fungus and Culture lines			
	Wheat* *E. graminis* f.sp. *tritici*			
	MR4c	MR4r	H1e	MB3e
AGE OF CULTURES				
Age of culture lines 3–32 weeks			+	+
63–95 weeks	+	+		
Time from subculture 0–6 weeks	+	+	+	+
MEDIUM				
± Sucrose (0–6%)	+	+	+	+
± Hormones (NAA, 2,4-D, BA, kinetin)	+	+	+	+
x 0.25 salts	+	+	+	+
x 0.50 salts	+	+	+	+
ENVIRONMENT				
r.h. 78.4, 88.3, 90, 100%				
Mannitol 0.3 M or 0.4 M				
25°C, continuous light	+	+	+	+
20°C, 16 h light	+	+	+	+
15°C, 16 h light	+	+	+	+
20°C, 16 h light; 15°C, 8 h dark	+	+	+	+
SPORE INOCULUM				
Dry	+	+	+	+
In FC 43 (fluorinated hydrocarbon)	+	+	+	+
INTERFACE OVERLAY				
Waxy layer	+	+	+	+
Enzymatically isolated pea cuticle				

*Cultures grown on medium of Murashige & Skoog (1962)
[†]Cultures grown on B5 medium (Gamborg *et al.*, 1968)
[¶]Cultures grown on medium of Heller (1953)
+Infection trial made

Table 1 (continued)

Pea[†] E. pisi PSO3	OP1	OP2	YP1	P2	PL1	PL3	PL4	Rose[¶] S. pannosa PSR	Sycamore[¶] U. aceris AM
+	+	+	+	+	+	+	+		
								+	+
+	+		+	+	+			+	+
+	+		+	+	+			+	+
+	+		+	+	+			+	+
+	+		+	+	+			+	+
+	+		+	+	+			+	+
+	+				+				
+	+				+				
+	+		+	+	+			+	+
+	+				+				
+	+				+				
+	+				+				
+	+	+	+	+	+	+	+	+	+
+	+	+	+	+	+	+	+	+	+
+	+				+				
+	+		+		+				

(a fluorinated hydrocarbon previously used for powdery mildew inoculation by Bushnell & Rowell, 1967). Examine cultures, using a binocular
microscope, daily for 2 weeks, discarding contaminated cultures.
Scanning electron microscopy may be used for more detailed observations.

Inoculation of tissue cultures with powdery mildew fungi

Plate suspension cultures of rose and sycamore by spreading the cells
onto, or incorporating them into, nutrient agar (1 ml cells plus 9 ml
medium) in a Petri dish. Transfer small pieces of callus (*c.* 2 mm^3) of
wheat and pea onto nutrient agar medium with the minimum of disturbance.
Seal the Petri dishes with Nescofilm and incubate under the appropriate
temperature and photoperiod for the desired length of time before
subjecting the cultures to any further experimental regime and inoculating. Alteration of the relative humidity (r.h.) of the cultures can be
achieved by incubating the unsealed Petri dishes over an appropriate
solution (water, 100% r.h.; saturated zinc sulphate, 90% r.h.) in a
sealed container. Leave to equilibrate before inoculating the cultures.

Table 1 summarizes the experimental inoculation regimes which we used
for each species, variety and culture line. None of the treatments
resulted in the development of mildew colonies. In all cases abnormal
conidial germination occurred, but the germination hyphae did not
produce appressoria. Some of these hyphae grew in close contact with
the cells whilst others grew away from them. The presence of the
overlaid isolated natural cuticle diminished the latter response.

Inoculation of detached pea leaves

Callus on detached leaves   Wound small areas of the lamina with a
needle and maintain on B$_5$ medium. When the callus is established
inoculate both the leaf epidermis and the callus with purified conidia.
In our experiments observation over a period of weeks revealed the
failure of mildew to establish colonies over the callus areas. Similar
results have been reported for rose by Mence & Hildebrandt (1966).

Restructured leaves   Using fine forceps, remove strips of the lower
epidermis of the leaves. Leaving some areas of the mesophyll exposed,
cover other areas with newly isolated waxy layers or natural cuticle.
Leave the leaves in unsealed Petri dishes containing water-agar with
1.0 mg/l BAP for 24 h, then inoculate the exposed mesophyll cells, the
mesophyll overlaid with isolated cuticle and the normal leaf (with or
without overlaid cuticle). In our experiments conidia did not germinate
normally: i.e. they did not produce appressoria on either the exposed
mesophyll or the mesophyll which had been overlaid with isolated cuticle.
However, the presence of isolated cuticle over the unaltered leaf
surface did not impede the growth of the colonies which had become
established on intact surfaces. Where isolated cuticle overlaid the
mesophyll and was also in contact with the normal leaf surface, the
fungal hyphae continued growth onto the synthesized surface. Clearing
the tissues with chloroform:lactic acid:methanol (1:1:1) revealed
occasional haustoria within these mesophyll cells. By contrast few

hyphae continued growth in contact with exposed mesophyll cells, no haustoria being observed in these cells.

DISCUSSION

In the trials described here no powdery mildew colonies were established on any of the culture lines of rose, sycamore, wheat or pea. Unsuccess-ful attempts to infect rose cultures have also been reported by Mence & Hildebrandt (1966). These workers cultured leaves of 12 rose varieties and attempted to infect the resultant callus both on the explant and after removal to fresh medium. Even on leaves infected prior to culture the fungus did not grow on the callus, nor did it produce haustoria in callus cells.

By contrast other workers have reported the successful establishment of mildew colonies on callus of *Vitis* spp. (Morel, 1948) and tumour tissues of *H. annuus* (Heim & Gries, 1953). The latter authors attempted to establish mildew colonies on both callus tissues and the slower growing undifferentiated tumour tissues derived from secondary crown-galls induced in *H. annuus* by *Agrobacterium tumefaciens*. Only inocula-tion using a dry sterile mounted needle resulted in infection, with about 2% of the tumour tissues inoculated supporting the limited growth and sporulation of the colonies.

The influence of the high humidity in the culture vessel (Morel, 1948) and the moisture on the tissue surfaces (Heim & Gries, 1953) on the growth of the colony has been considered in experiments. Morel (1948) moistened the callus surface prior to inoculation and transferred conidia to the wet tissue surface. The problem of high humidity in the culture vessel was overcome by leaving the vessels open. Under these conditions he obtained rapid mildew growth, but the callus grew slowly and became enveloped by the fungus. This technique necessitated an increase in the frequency of subculture and presumably also increased the incidence of contamination by unwanted fungi and bacteria. By contrast Heim & Gries (1953), using sealed vessels, claimed that the more vigorous tumour tissues (inoculated dry) supported powdery mildew growth, and drew attention to the fact that obligate parasites grow best on vigorous hosts. In neither investigation was the fungus from a dual culture used to reinoculate host plants.

In the experiments described here, using rose, sycamore, wheat and pea callus cultures, attempts were made to alter various factors which might influence the establishment of powdery mildew colonies. Where possible several culture lines of different morphological characteris-tics, age and growth rates were used. In addition variations were made in the physical environment (i.e. medium, temperature, photoperiod, relative humidity and the surface of the tissues) and in the method and time of inoculation of the cultures within the diurnal cycle and within the culture's growth cycle (i.e. during lag, exponential or stationary phases). Whilst none of these treatments resulted in mildew establish-ment, experiments with detached pea leaves provided a valuable indicator

of the interface necessary for infection. The failure to establish
mildew colonies on pea callus overlayed with natural isolated cuticle
may be related to the presence of the phytoalexin pisatin in cultured
pea tissues (Bailey, 1970). The use of either natural isolated cuticles
or synthetic membranes may prove valuable in attempts to culture other
powdery mildew species.

REFERENCES

Bailey J.A. (1970) Pisatin production by tissue cultures of *Pisum
    sativum* L. *Journal of General Microbiology* 61, 409-15.
Bayliss M.W. & Gould A.R. (1974) Studies on the growth in culture of
    plant cells. XVIII. Nuclear cytology of *Acer pseudoplatanus*
    suspension cultures. *Journal of Experimental Botany* 25, 772-83.
Bushnell W.R. & Rowell J.B. (1967) Fluorochemical liquid as a carrier
    for spores of *Erysiphe graminis* and *Puccinia graminis* (for inocula-
    tion experiments on cereals). *Plant Disease Reporter* 5, 447-8.
Gamborg O.L., Miller R.A. & Ojima K. (1968) Nutrient requirements of a
    suspension culture of soy bean root cells. *Experimental Cell
    Research* 50, 151-8.
Gil F. (1976) Ultrastructure and physiological properties of haustoria
    of powdery mildews and their host interfaces. *Ph.D. Thesis,
    University of London.*
Heim J.E. & Gries G.A. (1953) The culture of *Erysiphe cichoracearum* on
    sunflower tumour tissue. *Phytopathology* 43, 343-4.
Heller (1953) Recherches sur la nutrition minérale des tissus végétaux
    cultivés *in vitro*. *Thèse, Paris* and *Annales des sciences naturelles,
    Paris* 14, 1-223.
Jarvis W.R. (1964) Thermal and translocated induction of endophytic
    mycelium in two powdery mildews. *Nature* 203, 895.
Jarvis W.R. (1968) Induction of endophytism and changes in virulence in
    some powdery mildews. *Proceedings of the First International
    Congress of Plant Pathology*, p. 97.
Lamport D.T.A. & Northcote D.A. (1960) Hydroxyproline in primary cell
    walls of higher plants. *Nature* 188, 665-6.
Mence M.J. & Hildebrandt A.C. (1966) Resistance to powdery mildew in
    rose. *Annals of Applied Biology* 58, 309-20.
Morel G. (1948) Recherches sur la culture associée de parasites
    obligatoires et de tissus végétaux. *Annales des Épiphyties (série
    Pathologie Végétale)* 14, 1-112.
Murashige T. & Skoog F. (1962) A revised medium for rapid growth and
    bioassays with tobacco tissue cultures. *Physiologia Plantarum* 15,
    473-97.
Nash D.T. & Davies M.E. (1972) Some aspects of growth and metabolism of
    Paul's Scarlet rose cell suspensions. *Journal of Experimental Botany*
    23, 75-91.
O'Hara J.F. & Street H.E. (1978) Wheat callus culture: the initiation,
    growth and organogenesis of callus derived from various explant
    sources. *Annals of Botany* 42, 1029-38.
Salmon E.S. (1905) Further cultural experiments with 'biologic forms'
    of the *Erysiphaceae*. *Annals of Botany* 19, 125-48.

Salmon E.S. (1906) On endophytic adaptation shown by *Erysiphe graminis* DC. under cultural conditions. *Philosophical Transactions of the Royal Society. Series B* 198, 87-97.

Schnathorst W.C. (1959) Growth of *Erysiphe cichoracearum* on isolated lower epidermis and spongy mesophyll of lettuce. *Phytopathology* 49, 115-6.

Stuart R. & Street H.E. (1969) Studies on the growth in culture of plant cells. IV. The initiation of division in suspensions of stationary phase cells of *Acer pseudoplatanus* L. *Journal of Experimental Botany* 20, 556-71.

Yang S.L. & Ellingboe A.H. (1972) Cuticle layer as a determining factor for the formation of mature appressoria of *Erysiphe graminis* on wheat and barley. *Phytopathology* 62, 708-14.

The interaction of plant parasitic nematodes with excised root and tissue cultures

M.G.K. JONES

Welsh Plant Breeding Station,
Plas Gogerddan, Aberystwyth SY23 3EB

INTRODUCTION

Excised root and tissue cultures provide a more convenient environment than soil for maintaining populations of root-infecting nematodes, for studying nematode behaviour and for investigating the effects of infestation on host metabolism.

Nematodes that attack plants can be divided into migratory and sedentary types. Migratory nematodes usually puncture cells, withdraw their contents and kill the cells; they can feed at the plant surface (e.g. *Trichodorus* spp., *Tylenchorhynchus* spp.) or within the tissues (e.g. *Ditylenchus* spp., *Pratylenchus* spp.). These nematodes can be maintained routinely on suitable callus tissues. The sedentary types are endoparasites, (e.g. *Meloidogyne* spp., *Heterodera* spp., *Globodera* spp., *Rotylenchulus* spp.) and their relationships with host cells are more complicated, since they need to induce large volumes of cytoplasm (giant cells or syncytia), and cannot grow or reproduce without them. The sedentary nematodes can be cultured on excised roots in which giant cells or syncytia develop, but cannot as yet be maintained on completely undifferentiated callus tissues.

The main problem in culturing phytoparasitic nematodes is in sterilizing the original inoculum. A delicate balance is required, so that the contaminant micro-organisms are killed without injury to the nematodes. Usually standard procedures for initiation and maintenance of tissue and root cultures are suitable. Once the original nematode inoculum is established and growth and reproduction proceed, the nematode-host tissue complexes can be subcultured. A reasonable balance between nematode growth and reproduction and tissue growth is needed for successful long term maintenance of the association in culture.

Ingram D.S. & Helgeson J.P. (1980) *Tissue Culture Methods for Plant Pathologists*

SOME USES OF NEMATODE CULTURES

Nematodes that feed from and kill individual cells

Krusberg (1961) noted that difficulty in obtaining nematodes in quantity had greatly handicapped many studies with these plant parasites. Now, however, using monoxenic cultures, many species can be obtained in gram quantities.

The uses of these cultures include maintenance of pure nematode populations, studies of nematode biology, genetics and nutrition, propagation of masses of nematodes for study of their physiology and composition, and screening for nematicidal compounds. Using monoxenic cultures Bingefors & Bingefors (1976) have maintained a "nematode bank" of well defined nematode races and populations of *Ditylenchus dipsaci*, some for over 10 years, without changes in host specificity or virility. These have been used to supply plant breeders, mainly in Scandinavia, with defined nematode populations for breeding for resistance. Other nematodes that have been cultured in this way include *Ditylenchus* spp., *Aphelenchoides* spp., *Pratylenchus* spp. and *Tylenchorhynchus capitatus* (Krusberg, 1961; Webster & Lowe, 1966; Bingefors & Bingefors, 1976).

Nematodes that feed from modified host cells

Nematodes that feed from modified host cells, such as terminal root galls (e.g. *Xiphinema* spp.), or which induce the formation of giant cells or syncytia (e.g. *Meloidogyne* spp., *Heterodera* spp., *Globodera* spp., *Nacobbus* spp., *Rotylenchulus* spp.), can be maintained on excised root cultures or discs of storage tissues. Because of the more complicated nature of the associations cultures have been employed mainly as experimental tools rather than as substrates to multiply nematode numbers. Uses include filming of nematode behaviour, investigating the effects of plant growth regulators and feeding radioactive precursors to examine altered host metabolism.

For example, the work of Sandstedt & Schuster (1965, 1966) showed that callus developed when carrot storage discs cultured on a basal medium were infected with juveniles of *Meloidogyne incognita*. On excised tobacco pith, which required both cytokinin and auxin for callus production, neither hormone could be replaced by nematodes to maintain growth, but the presence of nematodes caused an increase in callus production. It was concluded that the nematodes secreted neither auxins nor cytokinins, but that nematode secretions trapped transported growth regulators at the infection sites. However, recent studies have shown that galls develop when excised roots of *Impatiens balsamina* are infected with *Meloidogyne javanica* (Jones, 1980). With reinfection of the tissues by new juveniles after the first life-cycle is completed, nematode callus is produced. This can be subcultured as an amorphous lump (although at the cellular level it contains vascular tissue, giant cells and callus-like parenchyma cells). Because there is no root along which polar transport of auxin could occur, it appears that the nematodes stimulate endogenous auxin production. Measurements of auxin

protectors support the latter hypothesis (Jones, 1980). Auxin protectors are substances which inhibit the peroxidase catalysed oxidation of
indole 3-ylacetic acid (IAA) by acting as an alternative substrate,
causing a lag before IAA oxidation, rather than by changing the
subsequent rate of IAA oxidation (Stonier, 1971). In Fig. 1 nematode
callus of *I. balsamina* contains significantly greater protector activity
than, for example, stem tissue (*I. balsamina*) or potato sprouts or
tuber. Increased auxin levels may be obtained by increased synthesis
or decreased degradation. The results shown in Fig. 1 suggest that
decreased degradation of auxin may be one reason for its increased
levels in nematode callus. Cyst-nematodes (*Heterodera* spp., *Globodera*
spp.) do not cause galling or the development of nematode callus.

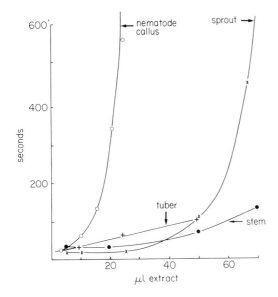

Figure 1. Auxin protectors assayed as described by Haard (1978).
The lag (in seconds) before peroxidase-catalysed oxidation of indole
3-ylacetic acid (IAA) occurs is a measure of auxin protectors present.
Similarly prepared extracts from nematode callus on *Impatiens
balsamina*, stem from *I. balsamina*, potato sprout and tubers were
added to the assay. Nematode callus extracts contain the most IAA
protector activity.

MATERIALS AND METHODS

Nematode sterilization

For successful results it is advisable to start with clean preparations
of active nematodes, freshly extracted from infested plants or freshly
hatched from egg sacs or cysts. If soil or other debris is present, a
preliminary cleaning is required. Large nematodes should be washed with

sterile water on a sieve of mesh size small enough to retain them. Small nematodes (especially endoparasites) should be washed through a fine mesh sieve (33-43 μm aperture) so that most of the debris is retained. An alternative procedure for preliminary cleaning is to collect nematodes after they have moved through sterile water agar.

Sodium hypochlorite should not be used as a sterilant as it dissolves nematode stylets.

Large nematodes (mainly migratory, such as Xiphinema *spp.,* Aphelenchoides *spp.,* Ditylenchus *spp.)*   These should be passed singly through solutions with micro-needles or mounted eye-lashes. Various treatments have been used successfully, including the following.
  1. Pass through 5-6 baths of a solution containing 20 mg/l malachite green and 1000 mg/l streptomycin sulphate for 3-4 h (Krusberg, 1961).
  2. Immerse in a solution containing 100 mg/l mercuric chloride for 2 min, and then in a 1% solution of streptomycin sulphate for a further 2 min (Dolliver et al., 1962).
  3. Immerse in a 0.03% solution of sodium azide for 90 min and then wash twice with sterile distilled water (Wyss, 1978).
See Hooper (1970) for further variations. A 0.5-1% solution of hibitane diacetate (ICI Ltd) may also be used (see below).

Small nematodes (sedentary juveniles and eggs)   These can usually be treated in large numbers by washing and sedimenting in capped conical centrifuge tubes (10 ml graduated) by centrifugation at 1000 rev/min for 2 min using a bench centrifuge. Glass tubes are preferable since nematodes often adhere to plastic tubes. Different procedures are used for sterilizing cyst-nematode eggs and juveniles, as follows.
  1. Treat cyst-nematode eggs with 20 volumes of hydrogen peroxide for 1 h.
  2. Treat juveniles with a 1% solution of hibitane diacetate for 25 min, then wash five times in sterile distilled water.
See Hooper (1970) for further variations.

Initiation of callus and root cultures

Callus of appropriate dicotyledonous hosts can be induced by standard techniques (Thomas & Davey, 1975; Helgeson, this volume), usually from surface sterilized seeds germinated on water or nutrient agar. Generally, conditions for good callus growth favour nematode reproduction. Media which have been used include those of Knop (modified Bingefors & Bingefors, 1976), Hildebrandt et al. (1946; modified Webster & Lowe, 1966), White (Sandstedt & Schuster, 1965), Murashige & Skoog (1962) and Gamborg & Eveleigh (1968; B5), with 2,4-dichlorophenoxyacetic acid (2, 4-D; c. 2 mg/l) as an auxin source, and solidified with agar (0.5-1.5%). Krusberg (1961) favours growth on slanted medium stored upright, since heavy infestations of D. dipsaci can cause some liquefication of the agar.

Root cultures are usually obtained from seeds surface sterilized by

immersion for 5 min in 100% ethanol and 25 min in a 50% commercial solution of sodium hypochlorite containing a drop of detergent, washed six
times in sterile distilled water and germinated on water or nutrient
agar (without hormones). Alternatively aseptic stem or bud explants may
be used. The medium of Gamborg & Eveleigh (1968; PRL-4, 1% agar) has
been used routinely to culture balsam, potato and tomato roots; that of
Murashige & Skoog (1962) has also been used (Jones, unpublished data).
Root cultures grow better in sealed Petri dishes than in tubes.

Wyss (1978) suggests a different approach for culturing the large
Xiphinema nematodes. Sterile fig seedlings germinated on 0.7% water
agar are kept in low light intensity, and the roots supplied with drops
of sterile Hoagland's solution. This reduces the effects of contamination, and provides a means of controlling culture growth.

Inoculation and subculturing

Nematode inocula can become trapped in surface water films, and these
should be avoided.

Inoculation levels can be varied as required. Krusberg (1961)
inoculated callus cultures with about 50 nematodes (*D. dipsaci*, *Aphelenchoides ritzemabosi*) and obtained 40 000 - 80 000 individuals after
2 months of culture at 22°C. Subsequent subcultures should be made by
inoculation of fresh callus cultures with small pieces of infected
callus containing 1000-3000 individuals (Bingefors & Bingefors, 1976).

Root cultures may be inoculated with one nematode if required, placed
on the agar near the root tip. For routine inoculations of endoparasites, about 25 juveniles should be added in a droplet to each root tip.
When accessible, egg sacs from established cultures of root-knot
nematodes provide a ready source of sterile juveniles; they can also be
used for direct infection of root cultures. For reproduction by cyst-
nematodes higher numbers are required, since males must leave the roots
and locate females, but this is not a problem for genera in which
reproduction is parthenogenetic. The agar surface may be sprinkled with
sterile sand to aid nematode penetration and to decrease surface tension
problems.

To subculture root cultures, infected portions should be excised and
transferred to fresh medium; new roots usually grow, but if such growth
is limited the cultures may be supplemented with fresh non-infested
roots.

Synchronous growth of the root and the sedentary parasite after the
initial inoculation does not occur in every culture, and it is advisable
to maintain enough cultures to allow for some culture losses.

REFERENCES

Bingefors S. & Bingefors S. (1976) Rearing stem nematode inoculum for plant breeding purposes. *Swedish Journal of Agricultural Research* 6, 13-7.

Dolliver J.S., Hildebrandt A.C. & Riker A.J. (1962) Studies of reproduction of *Aphelenchoides ritzemabosi* (Schwartz) on plant tissues in culture. *Nematologica* 7, 294-300.

Gamborg O.L. & Eveleigh D.E. (1968) Culture methods and detection of glucanases in suspension cultures of wheat and barley. *Canadian Journal of Biochemistry* 46, 417-21.

Haard N.F. (1978) Isolation and partial characterization of auxin protectors from *Synchytrium endobioticum* incited tumors in potato. *Physiological Plant Pathology* 13, 223-32.

Hildebrandt A.C., Riker A.J. & Duggar B.M. (1946) The influence of the composition of the medium on growth *in vitro* of excised tobacco and sunflower tissue cultures. *American Journal of Botany* 33, 591-7.

Hooper D.J. (1970) Laboratory methods for work with plant and soil nematodes. *Ministry of Agriculture, Fisheries and Food Technical Bulletin Number 2*, pp. 96-114. Her Majesty's Stationery Office, London.

Jones M.G.K. (1980) Host cell responses to endoparasitic nematode attack: structure and function of giant cells and syncytia. *Annals of Applied Biology* (in press).

Krusberg L.R. (1961) Studies on the culturing and parasitism of plant parasitic nematodes, in particular *Ditylenchus dipsaci* and *Aphelen-choides ritzemabosi* on alfalfa tissues. *Nematologica* 6, 181-200.

Murashige T. & Skoog F. (1962) A revised medium for rapid growth and bioassays with tobacco tissue cultures. *Physiologia Plantarum* 15, 473-97.

Sandstedt R. & Schuster M.L. (1965) Host-parasite interaction in root-knot nematode-infected carrot tissue. *Phytopathology* 55, 393-5.

Sandstedt R. & Schuster M.L. (1966) The role of auxins in root-knot nematode-induced growth on excised tobacco stem segments. *Physio-logia Plantarum* 19, 960-7.

Stonier T. (1971) The role of auxin protectors in autonomous growth. *Les Cultures de Tissus de Plantes*, pp. 423-35. Colloques Inter-nationaux du Conseil Nationale de Recherche Scientifique, Number 193, Paris.

Thomas E. & Davey M.R. (1975) *From Single Cells to Plants*. Wykeham Publications Ltd, London.

Webster J.M. & Lowe D. (1966) The effect of the synthetic plant growth substance, 2,4-dichlorophenoxyacetic acid, on the host-parasite relationships of some plant-parasitic nematodes on monoxenic callus cultures. *Parasitology* 56, 313-22.

Wyss U. (1978) Root and cell response to feeding by *Xiphinema* index. *Nematologica* 24, 159-66.

Vesicular-arbuscular mycorrhiza in root organ cultures

C.M. HEPPER & B. MOSSE
Rothamsted Experimental Station, Harpenden,
Hertfordshire AL5 2JQ

INTRODUCTION

Vesicular-arbuscular (VA) mycorrhizal fungi are obligate symbionts and
have not, so far, been grown on a synthetic medium without a host plant.
Our work on the development of a technique for producing mycorrhizal
infections in root organ cultures sought to answer three questions:
whether mycorrhizal fungi could infect a root which had no attached
shoot system; whether infected root organ cultures could be an intermediate stage in the progress towards the culture of these fungi; and
whether the system could be refined to produce axenic material in liquid
culture for biochemical studies.

We found that typical infections could develop in root organ cultures
of *Trifolium pratense* even after these had been maintained by subculturing for up to 3 years. This discounted the possibility that any
residual material from the shoot was necessary for the initiation or
development of the symbiosis.

The medium used to grow the root organ cultures contains 2% sucrose.
Since previous evidence suggested that such sugar concentrations might
depress the growth of mycorrhizal fungi (Mosse, 1959) we tested the
effect of sugar on the growth of the fungus using divided Petri dishes,
which allow two different media to be used simultaneously. When the
proximal end of the root was in a medium containing sucrose and the
fungus was only in contact with inorganic nutrients, mycorrhizal
development was no better than when the inoculum was placed on complete
medium.

By analogy with the culture of *Gymnosporangium* spp. (Cutter, 1959),
which developed by outgrowth from infected callus cells, it seemed that
root organ cultures might behave similarly and act as an intermediate
stage in axenic culture of the endophyte. So far, however, they have
not been used successfully to initiate cultures of the fungus alone.
We also attempted, using calluses derived from tobacco shoots and from
roots of soyabean, lima bean and carrot, to infect callus cultures with
VA mycorrhizal endophytes, as a stage towards axenic culture, but were

Ingram D.S. & Helgeson J.P. (1980) *Tissue Culture Methods for Plant Pathologists*

unsuccessful.

Although good infections developed in root organ cultures of *T. prat-ense*, infection failed to keep up with root growth when infected roots were subcultured. Root organ cultures in liquid medium were less well infected than those on agar media and we were unable to use this system to maintain a supply of axenic fungal material for biochemical studies.

Since *T. pratense* presents some problems for root organ culture work, other host plants should be tried.

MATERIALS AND METHODS

Host

Root organ cultures are initiated from axenic seedlings using standard techniques (e.g. White, 1954). In the case of red clover (*T. pratense*) the seeds are sterilized by immersion in concentrated sulphuric acid for 20 min followed by 10 washes in sterile water. They are then allowed to germinate in inverted Petri dishes containing 1% water agar. The radicals are excised before they touch the lid of the dish and trans-ferred, in groups of 10, to 150 ml conical flasks containing 50 ml of medium. Single root tips of red clover usually fail to grow when placed in this volume of medium. At no stage in this procedure should the roots be allowed to dry out.

The medium used contains (in 1 l of distilled water): KCl, 65 mg; KNO_3, 80 mg; $Ca(NO_3)_2.4H_2O$, 300 mg; $MgSO_4.7H_2O$, 720 mg; $NaH_2PO_4.2H_2O$, 10.7 mg; FeNaEDTA, 4.6 mg; $MnCl_2.4H_2O$, 4.9 mg; KI, 0.75 mg; H_3BO_3, 1.5 mg; $ZnSO_4.H_2O$, 1.9 mg; $CuSO_4.5H_2O$, 1 µg; $Na_2MoO_4.2H_2O$, 0.17 µg; glycine, 3 mg; thiamine HCl, 0.1 mg; nicotinic acid, 0.5 mg; pyridoxine, 0.1 mg; sucrose, 20 g. The pH is adjusted to 4.9 and the medium autoclaved at 121°C for 10 min.

Roots are removed from the flasks and subcultured into fresh medium at 10 day intervals. Pieces of healthy root are cut either from the apical portion of the main axis or from further back along the root where secondaries are well developed. Unlike other species, *T. pratense* cannot be maintained indefinitely by subculturing exclusively from the tip of the main axis.

Fungus

VA mycorrhizal fungi are maintained on host plants in soils which are low in nitrogen and available phosphorus and which have been sterilized to eliminate indigenous endophytes (Mosse, 1973).

Resting spores are extracted from these infested soils by passing a soil suspension through a series of graded sieves (Gerdemann, 1955). The choice of mesh size depends on the endophyte; we have found 700, 250 and 106 µm the most suitable, the first one for retaining roots and

larger pieces of organic material and the other two for spores and
sporocarps of different sizes.

The material retained on the 250 and 106 μm sieves is suspended in
water and stored at 2-6°C for at least 6 weeks before use, since this
treatment usually improves the rate of germination. In the case of
Glomus mosseae (Gerdemann & Trappe, 1974: classification of the yellow
vacuolate spore type) it is best to pick up the individual sporocarps
under a microscope and place them on damp filter paper for storage.
The spores are then dissected out from the sporocarp with a needle
immediately before use. Non-sporocarpic spores are collected by
pipetting a suspension of the material retained on the sieves onto
filter paper; individual spores can then be collected under a microscope with a micropipette.

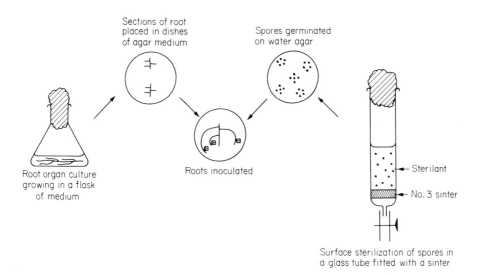

Figure 1. Diagram illustrating the steps involved in setting up
mycorrhizal root organ cultures. The maintenance of root organ
cultures is shown on the left, and the sterilization and germination
of fungal spores on the right.

Spores are surface sterilized in a solution containing chloramine T
(20 g/l), streptomycin (200 mg/l) and a trace amount of Tween 80. They
are left in sterilant for 20 min and then washed in three changes of
sterile distilled water. Non-sporocarpic spores can be surface
sterilized in watchglasses using a micropipette for transfers during
the washing procedure. Spores excised from sporocarps, however, have a
tendency to float and collect at the meniscus in watchglasses. A
convenient technique for sterilizing such spores using a pre-sterilized

glass tube fitted with a No. 3 sinter is illustrated in Fig. 1. The removal of the sterilant and washings from the apparatus is facilitated by inserting a syringe needle into the rubber tubing above the clip and drawing the liquid through. At no stage should the spores be drained completely dry.

The sterilized spores are placed in groups of 5-10 on 1% water agar. The choice of agar is very important since some types contain inhibitory levels of heavy metals or are unsuitable for other reasons (Hepper & Smith, 1976). We find Difco Bacto agar to be the most suitable. When maintained at 20-25°C spores should germinate within 5-10 days.

Axenically infected mycorrhizal roots can also be used as inoculum. When these are cut into short lengths, hyphae will grow from them and can infect another root placed close by. A method of obtaining such seedlings has been described by Mosse & Phillips (1971).

Establishment of dual cultures

Dual cultures are established on the same medium as root organ cultures, modified as follows: the pH is adjusted to 7.0 and the phosphate is supplied either as KH_2PO_4 (9.3 mg/l) or from a slow release source such as calcium phytate (630 mg/l). All media are solidified by the addition of 0.8% agar and sterilized at 121°C for 10 min.

Pieces of main root, approximately 20 mm long and bearing several laterals, are placed on the modified agar medium (2 root pieces/Petri dish). After 2-3 days they are inoculated by placing groups of pregerminated spores close to actively growing lateral root tips. The spores are best transferred on a small piece of agar using a sterile spatula. Plates are incubated in the dark at 20-25°C for up to 28 days.

Observations can be made at any time by viewing the dishes from the underside with a low power microscope. If a sterile work area is available the plates may be opened. At the end of the experiment roots are stained by being plunged into boiling lactophenol containing 0.05% cotton blue. They are left to simmer for 1 min, destained in lacto-phenol and viewed in lactic acid. Infections can also be observed in sections of root cut on a freezing microtome and stained with cotton blue.

Details of the prepenetration growth from the fungal spores and the nature and extent of VA mycorrhizal infection in cultured roots have been illustrated in earlier publications (Hepper & Mosse, 1975; Mosse & Hepper, 1975).

REFERENCES

Cutter V.M. (1959) Studies on the isolation and growth of plant rusts in host tissue cultures and upon synthetic media. I. *Gymnosporangium. Mycologia* 51, 248-95.

Gerdemann J.W. (1955) Relation of a large soil-borne spore to phyco-
  mycetous mycorrhizal infections. *Mycologia* 47, 619-32.
Gerdemann J.W. & Trappe J.M. (1974) The Endogonaceae in the Pacific
  Northwest. *Mycologia Memoir Number 5.*
Hepper C.M. & Mosse B. (1975) Techniques used to study the interaction
  between *Endogone* and plant roots. *Endomycorrhizas* (Ed. by
  F.E. Sanders, B. Mosse & P.B. Tinker), pp. 65-75. Academic Press,
  London.
Hepper C.M. & Smith G.A. (1976) Observations on the germination of
  Endogone spores. *Transactions of the British Mycological Society* 66,
  189-94.
Mosse B. (1959) The regular germination of resting spores and some
  observations on the growth requirements of an *Endogone* sp. causing
  vesicular-arbuscular mycorrhiza. *Transactions of the British
  Mycological Society* 42, 273-86.
Mosse B. (1973) Advances in the study of vesicular-arbuscular mycorrhiza. *Annual Review of Phytopathology* 11, 171-96.
Mosse B. & Hepper C.M. (1975) Vesicular-arbuscular mycorrhizal
  infections in root organ cultures. *Physiological Plant Pathology* 5,
  215-23.
Mosse B. & Phillips J.M. (1971) The influence of phosphate and other
  nutrients on the development of vesicular-arbuscular mycorrhiza in
  culture. *Journal of General Microbiology* 69, 157-66.
White P.R. (1954) *The Cultivation of Animal and Plant Cells*, 2nd
  edition. Ronald Press, New York.

Aseptic synthesis of sheathing (ecto-) mycorrhizas

P.A. MASON

Institute of Terrestrial Ecology, Bush Estate,
Penicuik, Midlothian EH26 0QB

INTRODUCTION

Some fungi, such as *Cenococcum geophilum*, form mycorrhizas with a large variety of trees, shrubs and even herbs, whereas others appear to be host specific. This physiological and ecological diversity offers excellent opportunities for selecting fungal species and strains that would be most suitable for field inoculation. We have therefore developed a simple and speedy screening procedure, which enables mycorrhizal development to be observed without destructive sampling.

The method is similar to those used for the study of legume nodule bacteria (Gibson, 1963) and root pathogens (Ebben & Last, 1973). Plants and fungi are grown together on defined agar media in transparent containers; unlike the the low nutrient regimes suggested by Pachlewska (1968) and others, the recommended balance of nutrients has enabled the rapid synthesis, within 4-6 weeks, of mycorrhizas closely resembling those found in nature. Mycorrhizas have been successfully synthesized on the roots of birch (*Betula* spp.) using isolates of *Amanita muscaria*, *Laccaria laccata*, *Laccaria proxima*, several *Leccinum* spp. and *Paxillus involutus*. The method has shown that inherent factors within both host and fungus, as well as root and shoot development (Mason *et al.*, 1977), control the formation of effective associations between birch and *A. muscaria* (Mason, 1975).

Although this method of mycorrhizal synthesis has been used with success when working with birch, results with spruce (Thomas & Jackson, 1979) suggest that the nutrient concentrations in the medium may have to be modified for species other than birch to ensure the formation of mycorrhizas anatomically and functionally similar to those found in the field.

Whereas mycorrhizal associations between *Betula* spp. and *A. muscaria*, *P. involutus* and *Leccinum* spp. form readily on an agar medium it is sometimes necessary to resort to Vermiculite/peat for those fungi (e.g. *Hebeloma* spp.) which form poor mycorrhizas on agar. Thus a Vermiculite/peat method of mycorrhizal synthesis is also presented here; although

Ingram D.S. & Helgeson J.P. (1980) *Tissue Culture Methods for Plant Pathologists*

possessing disadvantages it has been successfully used by a number of workers with a wider range of plant and fungal species than those mentioned above.

METHODS AND MATERIALS

1. *Production of contaminant-free birch seedlings*

   a. Collect birch seed from carefully noted locations between August and October. To retain their viability the seeds should be dried at room temperature for 2 days, placed in airtight bags and stored at 4°C.
   b. Place 50 seeds in a McCartney bottle (28 ml) and add 3 ml hydrogen peroxide (100 vol., 30% w/v).
   c. Place the bottle on a horizontal shaker and shake gently for 30 min. Afterwards ensure that the seeds remain in the hydrogen peroxide.
   d. Aseptically add sterile distilled water to float the seeds to the top of the bottle and transfer with a sterile needle to Petri plates of water agar (10 g agar/litre distilled water). Although the number of seeds per plate can be adjusted to suit experimental requirements 50 is optimal for most birch seedlots.
   e. For germination incubate the seeded plates for *c*. 10 days in continuous light, using Gro-lux (Thorn Ltd) or warm white fluorescent tubes, at *c*. 20°C (room temperature).
   f. During germination check the seeds/seedlings daily for contaminants. Because of their sparse growth on water agar these are sometimes difficult to detect. Remove contaminated seeds aseptically with a scalpel.
   g. After 10-14 days transfer fully emerged birch seedlings to tubes with a sterilized tungsten needle (see section 3.).

2. *Preparation of fungal inoculum*

   a. Sheathing mycorrhizal fungi that can be cultured are usually readily isolated in the autumn from tissue taken from their distinctive fruiting bodies, including toadstools, earth fans, elfin cups and truffles. Because of the risk of bacterial contamination take isolates from young fruitbodies, preferably before the veil of toadstools has broken. Air dry older specimens and retain for reference.
   b. Dry fruitbodies and flame very gently. Bend the caps and snap into two, exposing the cortical tissues. Remove small fragments of cap cortex with a sterile needle or scalpel.
   c. Place four fragments of cap cortex on the dried surface of Hagem's, modified Melin-Norkrans' (Table 1) or potato dextrose agar and incubate at room temperature. New fungal hyphae appear within 48-72 h, and colonies of *A. muscaria*, *L. laccata*, *L. proxima*, *P. involutus* and *Leccinum* spp. reach 2-3 cm diam. after 8 weeks. Care should be taken to remove contaminants as soon as possible.

Table 1. Media used for the production of birch mycorrhizas.

Hagem's medium for culturing mycorrhizal fungi

$MgSO_4.7H_2O$	0.5 g
KH_2PO_4	0.5 g
NH_4Cl	0.5 g
$FeCl_3$ (1% soln.)	0.5 ml
Glucose	5.0 g
Malt extract	5.0 g
Thiamine HCl	50 µg
Agar	10.0 g
Distilled water to 1 l	

Modified Melin-Norkrans' medium for culturing mycorrhizal fungi and synthesizing mycorrhizas

$CaCl_2$	0.05 g
NaCl	0.025 g
KH_2PO_4	0.5 g
$(NH_4)_2HPO_4$	0.25 g
$MgSO_4.7H_2O$	0.15 g
$FeCl_3$ (1% soln.)	1.2 ml
Thiamine HCl	100 µg
Glucose*	10.0 g
Agar	10.0 g
Distilled water to 1 l	

---

*Glucose is used rather than the sucrose (10 g) and malt extract
(3 g) specified by Marx (1969).

3. *Tube agar culture*

  a. Use sterile transparent polystyrene faeces bottles (Code No. 128A;
     Sterilin Ltd, Teddington, Middlesex). To enable gaseous diffusion,
     pierce a hole in the central channel of the top of each one with
     a hot needle and plug with cotton wool (see Pelham & Mason, 1978).
     After this modification replace the tops on the tubes and
     sterilize by cobalt irradiation. These containers can be re-used
     many times in this manner, but they should not be autoclaved.
  b. Prepare the three component solutions (A, B and C) of the modified
     Ingestad's medium (Table 2). Mix the solutions and add glucose,
     thiamine hydrochloride and agar.
  c. Autoclave the medium at 121°C for 15 min, aseptically pour 15 ml
     into each tube and slope to produce agar slants.
  d. Transfer one fully emerged birch seedling to each tube, placing it
     on, not into, the agar c. 2 cm from the top of the slope.
  e. Using a sterile scalpel, remove a 2-3 mm^3 block of inoculum from

the edge of a fungal colony and place at the tip of, or for a fast growing fungus such as *P. involutus* slightly (3-5 mm) in advance of, developing roots. Inoculate 5-10 replicate seedlings for each host/fungus combination.

f. Incubate the tubes at *c.* 20°C (room temperature) under a battery of Gro-lux lamps (continuous lighting).

g. Inspect the tubes regularly for possible contaminants. Birch seedlings should grow one leaf per week. As many fungi keep pace with the extending networks of roots, pieces of inoculum not fluffing out after 7-10 days should be replaced.

h. Mycorrhizas of *A. muscaria, P. involutus* and *Laccaria* spp. appear 4-6 weeks after inoculation. Their occurrence may be obscured by overlying wefts of fungal mycelium. After 8 weeks harvest the seedlings and inspect the mycorrhizas.

Table 2. Ingestad's medium for birch mycorrhizal synthesis (Mason, 1975; Pelham & Mason, 1978).

Prepare three solutions:

A	NH_4NO_3	106.2 g
	KNO_3	37.2 g
	K_2SO_4	22.2 g
	Distilled water to 1 l	
B	HNO_3	1.6 g
	$Ca(NO_3)_2$	14.3 g
	$Mg(NO_3)_2.6H_2O$	45.0 g
	$FeC_6H_5O_7.5H_2O$	2.094 g
	$MnSO_4.4H_2O$	0.811 g
	H_3BO_4	0.570 g
	$CuCl_2.2H_2O$	0.041 g
	$ZnSO_4.7H_2O$	0.054 g
	$Na_2MoO_4.2H_2O$	0.008 g
	Distilled water to 1 l	
C	KH_2PO_4	7.15 g/l (= 6.5 parts phosphorus/10^6)
	Distilled water to 1 l	

Mix 4 ml each of solutions A, B and C. Add glucose (10 g), thiamine hydrochloride (50 µg) and agar (10 g) and make up to 1 l with distilled water.

---

4. *Alternative procedure using Vermiculite/peat as a substrate*

a. Vermiculite granules can be bought in bulk from heating specialists.

Mix granules retained by a 2 mm sieve with acid peat at *c.* pH 3.5
to give, after adding nutrients and autoclaving, a pH of 4.8-5.2.
The ratio of Vermiculite to peat varies from batch to batch but is
near 9:1 by volume. Unlike Vermiculite, the peat used has passed
through a 2 mm sieve.
b. Place 250 ml of Vermiculite/peat mixture in each 500 ml wide-
necked Erlenmeyer flask, add 180 ml of modified Melin-Norkrans'
solution (Table 1) and autoclave for 30 min at 121°C.
c. Cool the mixture and inoculate each flask with four 2-3 mm^3 agar
blocks of inoculum.
d. Incubate the inoculated flasks at 20°C for 3-4 months. Inspect
regularly for possible contaminants and/or poor mycelial growth.
Shake the flasks each month to ensure thorough mixing and
permeation of the medium by the fungus.
e. Ten days before the inoculum is ready for use surface sterilized
seeds should be set to germinate (see *1.g.*).
f. When the inoculum and seedlings are ready for use transfer 15 ml
of the sterilized Vermiculite/peat mixture (see *4.b.*), with
nutrients but without fungus, to previously irradiated culture
tubes. Add one loopful of Vermiculite/peat fungal inoculum to
each tube, keeping it near the surface. Carefully place one
fully emerged birch seedling on the surface of the substrate in
each tube.
g. Incubate the tubes at 20°C under a battery of Gro-lux lamps
(continuous lighting) and keep under constant observation. Some
mycorrhizas will form against the walls of the tubes and be
apparent after 4-6 weeks. After 8-10 weeks wash the roots free
of Vermiculite/peat and record the results of the experiment.
h. Wide-necked Erlenmeyer flasks (500 ml) may be used instead of
culture tubes. In such cases prepare the flasks as described in
4.b. and inoculate the substrate medium with one tenth of its
volume of fully permeated inoculum (see *4.d.*). Incubate and
harvest as in *4.g.*, but note that experiments in flasks usually
take longer (10-12 weeks) than those in tubes (8-10 weeks).

REFERENCES

Ebben M.H. & Last F.T. (1973) Cucumber black rot caused by *Phomopsis
sclerotioides*. *Annals of Applied Biology* 73, 259-67.
Gibson A.H. (1963) Physical environment and symbiotic nitrogen fixa-
tion. 1. The effect of root temperature on recently nodulated
Trifolium subterraneum L. plants. *Australian Journal of Biological
Sciences* 16, 28-42.
Marx D.H. (1969) The influence of ectotrophic mycorrhizal fungi on the
resistance of pine roots to pathogenic infections. I. Antagonism
of mycorrhizal fungi to root pathogenic fungi and soil bacteria.
Phytopathology 59, 153-63.
Mason P.A. (1975) The genetics of mycorrhizal associations between
Amanita muscaria and *Betula verrucosa*. *The Development and Function
of Roots* (Ed. by J.G. Torrey & D.T. Clarkson), pp. 567-74. Academic
Press, New York and London.

Mason P.A., Pelham J. & Last F.T. (1977) Stimulation of anatomical changes in the stem of birch seedlings mycorrhizally infected with *Amanita muscaria*. *Nature* 265, 334-5.

Pachlewska J. (1968) Badania nad synteza mikoryzowa sosny (*Pinus sylvestris* L.) w czystych kulturach na agarze. *Prace 1BL* 345, 3-76.

Pelham J. & Mason P.A. (1978) Aseptic cultivation of sapling trees for studies of nutrient responses with particular reference to phosphate. *Annals of Applied Biology* 88, 415-9.

Thomas G.W. & Jackson R.M. (1979) Sheathing mycorrhizas of nursery grown *Picea sitchensis*. *Transactions of the British Mycological Society* 73 (1), 117-25.

Disease resistance studies with tissue cultures

J.P. HELGESON & G.T. HABERLACH

USDA, SEA, AR Plant Disease Resistance Research Unit,
Department of Plant Pathology, University of Wisconsin,
Madison, WI 53706, USA

INTRODUCTION

This article deals with the development of a tissue culture system for
the study of disease resistance. It includes aspects of the genetics
of host-pathogen interactions, the growth and maintenance of host and
pathogen, inoculation procedures, evaluation of reactions, physical
parameters and the effects of media variations on the interaction of
host and pathogen. Examples are drawn from our work with tobacco tissue
cultures and *Phytophthora parasitica* var. *nicotianae*, the cause of
black shank disease.

METHODS AND MATERIALS

Genetics of host and pathogen

If possible one should choose an experimental system for which the
genetics of both the host and the pathogen are well defined. It is
necessary to know whether discrete resistance genes exist and, if so,
whether they are dominant or recessive; whether the resistance genes are
single or multiple; whether resistant and susceptible host plants can
be crossed; whether more than one pathotype (race, pathovar) of the
pathogen exists and, if so, whether there are differences in virulence
and/or pathogenicity between the types. For example, in the tobacco-
P. parasitica var. *nicotianae* system resistance in the host is due to a
single dominant factor, effective against only one of the known races of
the fungus, and predictable behaviour of the F_1, F_2 and subsequent
crosses has been obtained. These attributes have proved very useful for
comparisons of plant and callus resistance (Helgeson *et al.*, 1976).

Growth and maintenance of host and pathogen

The host   The maintenance of healthy plants, good cultures and
virulent pathogens can be of critical importance for the experiments.
In particular, it is highly desirable for the tissue cultures and the
plants to be clonally related. With tobacco this can be achieved by

Ingram D.S. & Helgeson J.P. (1980) *Tissue Culture Methods for Plant Pathologists*

the following procedure. Remove apical buds from plants that are about 1 m tall. Wait 2-3 weeks and then excise the elongated axillary buds. Isolate pith tissue from the remaining stem and place in culture as below. Dip the base of the excised axillary buds in a proprietary rooting powder or solution (a mixture of an auxin and a fungicide) and plant in shallow pans containing a rooting mix such as expanded mica. Protect the cuttings for 2 weeks by covering the plant with a poly-ethylene bag and then uncover for about 2 weeks prior to inoculation or transplanting into pots with soil.

Growth of the tissue cultures will vary with the species and with the particular study desired. With phytoalexin elicitors it may be desirable to use suspension cultures (Dixon, this volume); at other times maintenance on agar medium may be more desirable. For the latter, Petri dishes are particularly convenient for growing tissue and applying inoculum. The procedures with tobacco are as follows. Cut a tobacco stem into 3-5 cm sections, sterilize the sections in a 1% solution of sodium hypochlorite for 15 min and then wash them twice with sterile, glass distilled water. Place the sections in sterile culture dishes and aseptically remove cylinders of pith with a sterile No. 1 or No. 2 cork borer. Extrude the pith sections into separate sterile culture dishes and cut each section into 2-3 mm discs. Place these discs on 50 ml of Murashige & Skoog (1962) medium supplemented with 11.5 µM indole 3-ylacetic acid (IAA; 2 mg/l) and 1 µM kinetin (0.215 mg/l) contained in 100 x 20 mm culture plates. Grow the tissues in the dark for 4 weeks at 26-28°C.

The pathogen   Growth conditions for the pathogen are also critically important. It must be maintained free of other micro-organisms and on a medium which gives reasonable yields of mycelium and infective units (zoospores, etc.). In addition, the maintenance of virulence of the fungus is of particular importance and this should be determined periodically, preferably with the clonal line of plants from which the susceptible cultures were obtained. One particular medium may be better than another for maintaining virulence. For *P. parasitica* var. *nicotianae* use an oatmeal-agar medium prepared by autoclaving 75 g of oatmeal and 20 g of agar in 1 l of distilled water. Inoculate Petri dishes containing 10-25 ml of this medium with a cube cut from a 4- to 5-week old plate. Zoospores are obtained after about 2 weeks of incubation.

Inoculation procedure

The collection and quantification of inoculum, the inoculation method and the concentration of inoculum are of critical importance. If possible inoculation should be with discrete entities such as zoospores; these can be counted with a haemocytometer, washed with buffers or water and then applied in appropriate dilutions. Inoculation with mycelium or mycelial fragments can be very difficult to repeat exactly. With several species of *Phytophthora* a cold treatment can stimulate zoo-sporangia formation and the subsequent synchronous release of zoospores. For *P. parasitica* var. *nicotianae* use a modification of the method

developed by Gooding & Lucas (1959), as follows. Pipette 10 ml of
water onto the mycelial mat in an oatmeal-agar plate, chill the plate
at 4°C for 30 min and then warm the plate to 22-24°C for 30-45 min.
Zoospores are released into the water and can be siphoned off with a
Pasteur pipette, counted and then diluted to the appropriate concentration. The mycelial mat is then stripped off the surface of the plate,
macerated and applied as a slurry-inoculum to the pans containing plants.

If possible, inoculate the cultured tissues without allowing contact
between the pathogen and the tissue culture medium, so that tissue and
pathogen interactions can be evaluated without providing an extra food
base for the pathogen. Rings (4 mm i.d. x 1-2 mm high) cut from plastic
tubing are satisfactory for retaining the pathogen. Place these on top
of the cultured tissues and introduce a measured volume of inoculum
(typically 30-50 µl) inside the ring. Use sterile water, buffers and
media in all operations so that the inoculum will contain only the
desired pathogen. Also, add sample drops of the inoculum to nutrient
broth to detect any bacterial contamination. After inoculation, return
the culture plates to the incubator, taking care not to dislodge the
rings and their contents.

In our early experiments we obtained the best results by inoculating
tissues with low numbers of zoospores (10-100 per tissue piece). Later,
when optimum conditions for the expression of resistance by the tobacco
tissues had been defined, it became possible to raise the zoospore
concentration by 100- to 1000-fold and still retain good resistance
expression. We recommend, therefore, that in the early stages of
development of a system the number of spores applied should be the
lowest number which will still assure contact with viable spores.

Evaluation of interactions

Within a few hours of inoculation tissues may begin to show a response
to the pathogen. With tobacco and *P. parasitica* var. *nicotianae* the
incompatible reaction is typically a hypersensitive response (HR) of the
host; this is evident by 6-7 h and intense by 24 h after inoculation.
Susceptible tissues show little or no HR when inoculated with living
spores. Observations within the first day or two after inoculation
often suggest the course of further interaction. By the third or fourth
day after inoculation tissues may show macroscopic evidence of infection.
In susceptible interactions this may result in a complete covering of
the tissue pieces by mycelium within 7-10 days. With tobacco, histo-
logical studies show differences in the degree of penetration of
resistant and susceptible tissues. After 48 h the penetration of the
tissue of a resistant callus is limited to only 5-8 cells, whereas 40-50
cell layers have been penetrated in the susceptible tissues (Gaard,
deZoeten & Helgeson, unpublished data).

A useful procedure for evaluation of infection is to determine the
changes in tissues of known genotypes and then to devise a rating scheme.
For example, with tobacco and *P. parasitica* var. *nicotianae* ratings are
0 if the tissues show no visible fungus, 1 if some fungal mycelium is on

the medium surrounding the tissues but not on the tissues themselves,
2 if aerial mycelium is seen on the tissues but not completely covering
them and 3 if the tissues are completely covered with mycelium (Helgeson
et al., 1976).

   It may also be useful to rate the tissues on a blind basis. For this,
have a colleague mark the cultures with code numbers so that ratings are
done without knowledge of the genotype of the source plant and fungus.
Then, starting 3 or 4 days after inoculation, examine each plate at 1-
or 2-day intervals and numerically rate the tissues. At the end of
the experiment make a subjective evaluation of the resistance or
susceptibility of the tissues before obtaining the key to the culture
code from your colleague.

   Another type of evaluation is to determine if a biochemical response
of the cultured tissue to a pathogen (e.g. phytoalexin production)
resembles that of the intact plant. Phytoalexins may be produced in
tissue cultures as well as in intact plants (Helgeson *et al.*, 1978;
Dixon, this volume). For example, tobacco leaves produce capsidiol in
response to tobacco necrosis virus (Bailey & Burden, 1973) or *P. parasitica* var. *nicotianae* (Tooley & Helgeson, unpublished data); tobacco
tissue cultures also produce this phytoalexin and at least three
additional closely related terpenoid phytoalexins (Helgeson *et al.*,
1978).

   Phytoalexins increase after challenge by fungi, and then inhibit
further fungal growth (Kuc, 1976). The latter property may be exploited
to detect phytoalexins on TLC plates (Bailey & Burden, 1973). Spot thin
layer plates (Silica gel G, Brinkman) with ethanol extracts of tissues,
run the plates with a suitable solvent and allow the solvent to
evaporate. Spray the plates lightly with nutrient agar and with a thin
layer of spores from a culture of *Cladosporium herbarum*, and incubate
for 2-3 days at 20-22°C in a chamber lined with moist paper. Inhibitory
zones appear as white areas of silica gel showing through a black fungal
background. Note that the volume of the medium and the spore suspension
should be minimized to avoid spreading of the spots. Also, the plates
should not be put in an upright position until the excess moisture has
evaporated.

Environmental conditions

The conditions for obtaining resistance expression in cultured tissues
may vary with the host and with the pathogen. With the *P. parasitica*
var. *nicotianae*-tobacco system 20°C appears to give the best difference
between resistant and susceptible genotypes. At 28°C the differential
breaks down and the tissues with the genetic factor for resistance to
race 0 are colonized by race 0 (Helgeson *et al.*, 1972). Conversely, at
16°C the tobacco tissues show damage due to the cool temperature, and
the fungus grows poorly on oatmeal-agar and on tissues of both genotypes.
A series of five or six batches of culture plates, each placed in an
incubator at a different temperature, will provide a test of the most
appropriate temperature for incubation. Because of the large heat

capacity of the water of the medium care should be taken to equilibrate the plates to the test temperature prior to inoculation. Overnight incubation at the inoculation temperature will suffice; then, if the plates are removed in small batches, inoculated immediately and returned without delay to the incubator the temperature inside the plates will remain constant.

If lighted incubators are available, determine if a specific light regime is required for resistance. With the *Phytophthora* species this may not be critical.

Media addenda

Various phytohormones are useful or even essential for controlling the growth of tissues in culture. For example, with tobacco loose friable tissues, tight hemispherical pieces or even buds on callus may be obtained simply by modifying the ratio of cytokinin to auxin in the culture medium (Helgeson, 1979). It is possible that only certain tissues will express resistance to a given pathogen; with *P. parasitica* var. *nicotianae* loose friable tissues are susceptible regardless of genotype, whereas the tight hemispherical tissues or rooted shoots from callus tissues are resistant if they have been initiated from a resistant plant.

Even if the tissue form is not changed by the phytohormone regime it is possible that the response to pathogens may vary. For example, tobacco tissues supplied with 11.5 µM IAA and either 10 µM kinetin or 10 µM benzyladenine are rapidly and completely colonized by race 0 of *P. parasitica* var. *nicotianae* even if the tissues have the resistance factor to this race (Haberlach *et al.*, 1978). In contrast, the same tissues will show genotype-specific resistance if they are supplied with 11.5 µM IAA and either 1 µM kinetin or 1 µM benzyladenine. The breakdown of resistance is associated with a lack of the usual hypersensitive response that accompanies resistance.

CONCLUSIONS

Ingram (this volume) has described the many aspects of the usefulness of plant tissue cultures to plant pathologists. In the area of disease resistance studies the experimental control obtained with tissue cultures can be extremely valuable: the cultures are grown without contaminating organisms; they can be maintained on defined media and give predictable responses to particular physiological regimes; in some cases they can be inoculated without wounding; the environmental control can be much more rigorous than in the field; and the number of different types of cells within the tissues can be considerably less than in organs of intact plants used for comparable studies. All these advantages are essential for a thorough investigation of disease resistance.

There are, however, some disadvantages of tissue culture systems for

resistance studies: protective tissues such as cuticle may be absent;
the tissues are often actively growing and therefore resemble only
meristematic areas of the plant; sometimes resistance is not expressed
in culture, or the biochemical events associated with resistance may
differ between plants and callus. All these problems are cause for
concern and must be studied carefully by anyone setting out to use
tissue cultures for disease resistance studies.

REFERENCES

Bailey J.A. & Burden R.S. (1973) Biochemical changes and phytoalexin
    accumulation in *Phaseolus vulgaris* following cellular browning caused
    by tobacco necrosis virus. *Physiological Plant Pathology* 3, 171-7.
Gooding G.V. & Lucas G.B. (1959) Factors influencing sporangial
    formation and zoospore activity in *Phytophthora parasitica* var.
    nicotianae. *Phytopathology* 49, 277-81.
Haberlach G.T., Budde A.D., Sequeira L. & Helgeson J.P. (1978)
    Modification of disease resistance of tobacco callus tissues by
    cytokinins. *Plant Physiology* 62, 522-5.
Helgeson J.P. (1979) Tissue and cell suspension culture. *Nicotiana:
    Procedures for Experimental Use. USDA Technical Bulletin* 1586 (Ed.
    by R.D. Durbin), pp. 52-9. US Government Printing Office, Washington.
Helgeson J.P., Budde A.D. & Haberlach G.T. (1978) Capsidiol: A
    phytoalexin produced by tobacco callus tissues. *Plant Physiology*
    61, Supplement p. 53.
Helgeson J.P., Haberlach G.T. & Upper C.D. (1976) A dominant gene
    conferring disease resistance to tobacco plants is expressed in
    tissue culture. *Phytopathology* 66, 91-6.
Helgeson J.P., Kemp J.D., Maxwell D.P. & Haberlach G.T. (1972) A tissue
    culture system for studying disease resistance: the black shank
    disease in tobacco callus cultures. *Phytopathology* 62, 1439-43.
Kuc J. (1976) Phytoalexins and the specificity of plant-parasite
    interactions. *Specificities in Plant Diseases* (Ed. by R.K.S. Wood
    & A. Graniti), pp. 253-63. Plenum Press, New York.
Murashige T. & Skoog F. (1962) A revised medium for rapid growth and
    bioassays with tobacco tissue cultures. *Physiologia Plantarum* 15,
    473-97.

Plant tissue culture methods in the study of phytoalexin induction

R.A. DIXON

Department of Biochemistry, Royal Holloway College
(University of London), Egham Hill, Egham, Surrey TW20 0EX

INTRODUCTION

Until recently the use of tissue culture techniques in plant pathology
was limited mainly to the establishment of dual cultures for the growth
of obligate parasites *in vitro* (Ingram & Tommerup, 1973). With the
increasing awareness of the role of phytoalexins (inducible, host-
synthesized antimicrobial agents) as important factors in plant disease
resistance there have been several investigations of the capacity of
plant cells to produce these compounds when grown in culture.

Plant tissue cultures can provide the research worker with aseptic
near-homogeneous cell populations growing under precisely defined
conditions, and such material is of undoubted value to the biochemist
for studies of phytoalexin biosynthesis and its enzymic regulation.
Such studies are important in view of our almost total lack of
knowledge of the enzymic reactions specific for the biosynthesis of
isoflavonoid or terpenoid phytoalexins.

Phytoalexins have now been detected in tissue cultures derived from
a number of plant species, although to date the work has mainly been
limited to members of the Leguminosae. The choice of plant species has
largely been dictated by the availability of information concerning
the extraction and quantitative determination of the phytoalexins,
and by the relative ease of culture of most leguminous plants. The
phytoalexins studied include pisatin from *Pisum sativum* (Bailey, 1970),
glyceollin from *Glycine max* (Ebel *et al.*, 1976), phaseollin from
Phaseolus vulgaris (Dixon & Fuller, 1976, 1978; Dixon & Bendall, 1978a),
medicarpin from *Canavalia ensiformis* (Gustine *et al.*, 1978) and
capsidiol from *Nicotiana tabacum* (Helgeson *et al.*, 1978). Although
terpenoid phytoalexin production in cultures of Solanaceous species
has so far received little attention, active resistance mechanisms
do appear to exist in such systems (Ingram & Robertson, 1965; Warren
& Routley, 1970; Helgeson *et al.*, 1976).

It is perhaps surprising that undifferentiated cell cultures appear
to retain their capacity for phytoalexin production, in view of the

Ingram D.S. & Helgeson J.P. (1980) *Tissue Culture Methods for Plant Pathologists*

generally held belief that de-differentiation is associated with a reduction in biosynthetic capability. It is often the case that restoration of secondary product biosynthesis only occurs following hormonally induced re-differentiation (Street, 1973). Hormone levels may, however, be an important factor in determining the extent of phytoalexin production in cultured cells (Dixon & Fuller, 1976, 1978).

The present article is intended to introduce the research worker to the basic methods required for studying phytoalexin induction in tissue cultures. The emphasis will be on isoflavonoid phytoalexins, although the general applicability of the tissue culture method and the non-specific nature of the majority of phytoalexin-inducing agents would suggest that the methods may also be suitable for use with species which produce phytoalexins of other structural classes.

MATERIALS AND METHODS

Media and techniques for the growth of callus and suspension cultures

It would appear from the limited literature available that leguminous plants will grow and still retain their capacity for phytoalexin induction in a variety of defined culture media. The media so far tested are shown in Table 1. There is at present no information available concerning the effects of different levels of salts, vitamins or carbon sources on phytoalexin production; most of the media listed in Table 1 are standard culture media formulated with cell growth as the main criterion. For many plant species no data are available on medium requirements, and for these the medium of Schenk & Hildebrandt (1972) is recommended, since this generally supports good growth of cultures from a wide range of monocotyledonous and dicotyledonous plants.

Full details of the standard methods for initiation and maintenance of callus and cell suspension cultures are given elsewhere (Street, 1977). The following procedure is used in our laboratory for the initiation of cultures of dwarf French bean (*P. vulgaris*). Seeds with intact testas are surface sterilized for 30 min with a sodium hypochlorite solution (1.0-1.4% available chlorine), washed and soaked overnight in sterile distilled water, and finally re-sterilized with the sodium hypochlorite solution and washed again with water. Radicle tips are then aseptically dissected from the seed and plated out in Petri dishes (five tips per dish) containing modified Schenk & Hildebrandt medium with the inclusion of 0.6% w/v agar (Dixon & Fuller, 1976). These are incubated in the dark at 25°C; enough callus usually forms in 5-6 weeks to allow subculture onto fresh medium. Calluses should then be maintained by regular subculture at 6 week intervals. With species which have very small seeds (e.g. lucerne or red clover) sterilized whole seeds rather than radicle tips should be plated out. These germinate rapidly and form calluses on roots and hypotocyls.

Suspension cultures are obtained by transfer of callus lumps to liquid medium and agitation on an orbital or reciprocal shaker. In the absence of information to the contrary, the medium used should be of the same composition as the medium for callus maintenance. Since many of the enzymes of isoflavonoid biosynthesis are light-induced (Grisebach & Hahlbrock, 1974), cultures for phytoalexin work are best grown in the dark or under low level illumination.

Extraction and estimation of phytoalexins from cultured tissues

Isoflavonoid phytoalexins may be extracted from callus and cell suspension cultured cells by methods identical to those used for their extraction from intact plants. In most methods cells are homogenized in aqueous ethanol, the ethanol removed under vacuum and the aqueous residue extracted with a non-polar organic solvent (usually ethyl acetate or petroleum ether). The organic phase is then concentrated under vacuum. Although most plant cell cultures do not contain chlorophyll, the presence of phenolic compounds necessitates a chromatographic step before quantitative determination is attempted. Thin-layer chromatography followed by elution and determination by ultraviolet spectrophotometry is the most widely used method (Dixon & Fuller, 1976; Gustine *et al.*, 1978; Hargreaves & Selby, 1978), although gas liquid chromatography (Bailey & Deverall, 1971) and, as shown more recently, high performance liquid chromatography (Zähringer *et al.*, 1979) are more sensitive methods for the detection and determination of low levels of phytoalexins.

In studies with cell suspension cultures, the levels of phytoalexins in the culture medium must also be determined. In French bean cell suspension cultures about 15% of the total phaseollin produced in response to treatment with denatured ribonuclease A accumulates in the culture medium (Dixon & Bendall, 1978a).

Constitutive phytoalexin production

Growth of plant cell suspension cultures is often slow during the first few passages after initiation, and during this period French bean cell cultures produce phytoalexin in the absence of added inducing agents (Dixon & Fuller, 1976; Hargreaves & Selby, 1978). This constitutive phytoalexin production, which does not occur in faster growing, established cultures, is possibly the result of cell death during the selection of faster growing cells in the suspensions. A similar phenomenon has been observed with pea callus cultures, which gradually lose their ability to produce pisatin in response to the coconut milk in the culture medium (Bailey, 1970). Whether these latter cultures retain their ability to produce phytoalexins in response to fungal infection was not, however, determined.

Constitutive phytoalexin production may be markedly affected by the levels of plant growth substances in the culture medium (Dixon & Fuller, 1976, 1978). The synthetic auxin 2,4-dichlorophenoxyacetic acid (2,4-D) is strongly inhibitory to phaseollin production at levels

Table 1. Growth media used in studies of phytoalexin induction and expression of disease resistance in plant cultures.

Species	Phytoalexin	Basic Medium*
Phaseolus vulgaris cv. Canadian Wonder (CW)	Phaseollin	Modified Schenk & Hildebrandt (SH)
Phaseolus vulgaris (CW)	Phaseollin	Modified SH
Phaseolus vulgaris cv. Kievitsboon koekoek	Phaseollin Phaseollidin Phaseollinisoflavan Kievitone	Murashige & Skoog
Glycine max cv. Harosoy & Harosoy 63	Glyceollin	Linsmaier & Skoog
Glycine max cv. Harosoy	Glyceollin	B_5
Canavalia ensiformis	Medicarpin	Modified Miller's medium
Pisum sativum	Pisatin	'Medium I' (+ 15% v/v coconut milk)
Solanum tuberosum	ND	See Ref. (includes 13% v/v coconut milk)
Nicotiana tabacum	ND	Linsmaier & Skoog
Nicotiana tabacum	ND	Linsmaier & Skoog
Nicotiana tabacum	Capsidiol	Linsmaier & Skoog[†]
Lycopersicon esculentum	ND	Murashige & Skoog (+ 15% v/v coconut milk)

* For basic media sources see references in final column.
[†] No information given on growth regulator levels.
ND, phytoalexin accumulation not investigated; 2,4-D, 2-4-dichloro-phenoxyacetic acid; NAA, 1-naphthaleneacetic acid; IAA, indole 3-ylacetic acid; BA, N^6-benzyladenine.

Table 1 (continued)

Growth regulator levels (mg/l)					Reference
2,4-D	NAA	IAA	Kinetin	BA	
Various	Various	-	Various	-	Dixon & Fuller, 1976, 1978
-	0.93	-	1.08	-	Dixon & Bendall, 1978a
0.5	-	-	0.1	-	Hargreaves & Selby, 1978
1.0	-	-	1.0	-	Keen & Horsch, 1972
2.0	-	-	-	-	Ebel *et al.*, 1976
-	0.4	-	0.25	-	Gustine *et al.*, 1978
6.0	-	-	-	-	Bailey, 1970
6.0	0.1	-	-	-	Ingram & Robertson, 1965
-	-	2	0.215	-	Helgeson *et al.*, 1972
0.11	-	-	0.215	-	Helgeson *et al.*, 1972
					Helgeson *et al.*, 1978
-	1.0	-	-	1.0	Warren & Routley, 1970

above 2×10^{-6} M, whereas substitution of 2,4-D (2×10^{-6} M) by
1-naphthaleneacetic acid (NAA; 2×10^{-6} M) leads to increased production. These effects may be a result of *in vivo* inhibition by 2,4-D of
the induction of L-phenylalanine ammonia-lyase, the enzyme catalysing
the first step of phenylpropanoid (and thus isoflavonoid) biosynthesis
(Davies, 1972).

   In French bean cell suspension cultures maximum rates of constitutive phytoalexin production are always less than 30% of the maximum
rates of synthesis in response to inducing agents. In view of this,
and the toxicity of phaseollin to cell suspension cultures (Glazener
& van Etten, 1978), it is clearly advantageous to use well-established
cultures for phytoalexin induction studies.

Induced phytoalexin production

One of the central problems in phytoalexin research is the apparent
lack of structural similarity between the many compounds of known
phytoalexin-inducing potential. Thus there is a wide choice of possible
inducing agents for use with cultured cells. Those used to date are
summarized in Table 2. Fungal spore suspensions are clearly the
preferred natural agents, and these may be used successfully to induce
phytoalexin accumulation in callus cultures (Keen & Horsch, 1972;
Gustine *et al.*, 1978). Spore suspensions are, however, generally
unsuitable for use with suspension cultures as the fungus may grow in
the culture medium without necessarily infecting the plant cells.
Alternative natural inducers for use with cell suspension cultures
are elicitor macromolecules obtained from fungal culture filtrates
or by heat treatment of fungal cell walls (Ebel *et al.*, 1976; Dixon
& Fuller, 1977; Dixon & Lamb, 1979). The wall-released elicitors
from *Colletotrichum lindemuthianum* and *Phytophthora megasperma* var.
sojae are potent inducers of phaseollin and glyceollin in suspension
cultures of French bean and soyabean respectively. Nigeran, a
commercially available (Koch-Light Laboratories Ltd) α-1,3,α-1,4
linked glucan from *Aspergillus niger* cell walls, has also been used
successfully as a phytoalexin-inducing agent in the soyabean system
(Ebel *et al.*, 1976), although it is much less effective with French
bean cultures.

   Early work on the phytoalexin response revealed that heavy metal
salts and certain antimetabolites are powerful inducers when applied
to endocarp, cotyledon or hypocotyl tissues of leguminous plants.
Their use with tissue culture systems has not been fully investigated,
although mercuric chloride and copper sulphate are very poor inducers
of phaseollin accumulation in French bean suspension cultures. It
is likely that these compounds act indirectly by causing the limited
death of cells in contact with the compound; phytoalexins then
accumulate in living cells at the periphery of the necrotic region
(Mansfield *et al.*, 1974). In suspension cultures consisting of single
cells and small clumps, all cells will readily come into contact with
heavy metal and thus be killed.

    Denatured ribonuclease and deoxyribonuclease are potent inducers
of phytoalexin accumulation, RNAase A (0.5 mg/ml; Sigma Chemical Co.,
London) causing greater phaseollin induction in French bean cell
suspension cultures than optimal concentrations of *Colletotrichum*
elicitor (Dixon & Lamb, 1979). As RNAase probably acts by non-specific
effects at the plant plasma membrane, it may also be an effective
inducing agent in other systems. Even more effective in inducing
phaseollin production in suspension cultures is the addition of an
extract from autoclaved bean hypocotyls (Hargreaves & Selby, 1978).
This preparation is believed to contain an internal endogenous elicitor,
which may play a role in the response of plant cells to exogenously
applied biotic or abiotic elicitors (Hargreaves & Bailey, 1978).

    The effects of plant growth regulators on the capacity for induced
phytoalexin production by cell cultures may be similar to their effects
on constitutive production (Dixon & Fuller, 1978). In French bean
cultures 2,4-D is inhibitory at high concentrations, while NAA (up to
at least 2×10^{-5} M) permits good rates of synthesis. The inhibition
of phaseollin induction by 2,4-D (2×10^{-6} M) is partially reversed by
increased levels of kinetin (up to 5×10^{-5} M). Effects of growth
regulator levels on other plant cultures, or on phytoalexin accumulation induced by different elicitors, have not been studied. The
problem for the research worker is to manipulate the components of the
culture medium in such a way as to obtain good growth rates and low
background rates of phytoalexin accumulation combined with high rates
of synthesis in response to added inducers. A brief but systematic
evaluation of the effects of auxin to cytokinin ratios may be the best
way to proceed if initial culture conditions prove unsuitable.

Criticisms and advantages of the cell culture approach

The major criticism of the application of tissue culture methods to
the study of phytoalexin induction concerns the problem of relating
results to the intact host-parasite system. This problem has been
well demonstrated with respect to the response of soyabean callus
cultures to *P. megasperma* var. *sojae* (Keen & Horsch, 1972), where
the extent of glyceollin production by cultures of different host
cultivars was quantitatively different from that of the intact plant.
However, it is also clear that race-specific resistance may be
wholly, or partially, expressed in culture (Warren & Routley, 1970;
Helgeson *et al.*, 1976), and in the latter case the phytoalexins
found in tobacco plants (Bailey *et al.*, 1975) are also found in
challenged callus. Further work is required to define more fully the
similarities and differences between cell cultures and their tissues
of origin with respect to the specificity of induced resistance
responses. This is especially necessary since cell suspension cultures
and plant protoplasts (derived either from cultures or isolated
mesophyll cells) provide ideal systems for studying the early
biochemical events associated with elicitor binding at the plant cell
surface, a phenomenon which may be an important determinant of race
specificity.

Table 2. Response of plant cell cultures to biotic and abiotic elicitors.

Elicitor	Concentration (per ml)	Plant species
BIOTIC ELICITORS		
Pithomyces chartarum spore suspension	10^7 spores	*Canavalia ensiformis*
Phytophthora megasperma var. *sojae* (PMS) spore suspension	10^5 spores "	*Glycine max* cv. Harosoy " " " Harosoy 63
PMS wall-released elicitor	1.0 µg	*Glycine max*
PMS wall-released elicitor	0.1 mg (CH_2O)	*Phaseolus vulgaris* cv. Canadian Wonder (CW)
Colletotrichum lindemuthianum wall-released elicitor	0.01 mg (CH_2O)	*Phaseolus vulgaris* (CW)
Botrytis cinerea solubilized wall preparation	0.1 mg (CH_2O)	*Phaseolus vulgaris* (CW)
Botrytis cinerea culture filtrate	100 µl	*Phaseolus vulgaris* (CW)
Nigeran (*Aspergillus niger* cell wall preparation)	0.5 mg	*Phaseolus vulgaris* (CW)
Nigeran (*Aspergillus niger* cell wall preparation)	0.04 mg	*Glycine max* cv. Harosoy 63
French bean hypocotyl extract	50 µl	*Phaseolus vulgaris* cv. Kievitsboon koekoek
ABIOTIC ELICITORS		
Mercuric chloride	0.86 mg	*Canavalia ensiformis*
Cupric sulphate	0.125 mg	*Phaseolus vulgaris* (CW)
Poly-L-lysine (av. mol. wt. 3400)	1.0 mg	*Phaseolus vulgaris* (CW)
Denatured RNAase A	0.5 mg	*Phaseolus vulgaris* (CW)

* C, callus culture; S, cell suspension culture.
† Amounts are expressed per g fresh weight of plant material.
¶ Amount per g dry weight.
** Phaseollin levels of 4 µg/g fresh weight are observed in the absence of elicitor.

Table 2 (continued)

Phytoalexin				Reference
Culture*	Compound	Amount[†] (μg)	Time (h)	
C	Medicarpin	60	48	Gustine *et al.*, 1978
C	Glyceollin	500–980[¶]	24–48	Keen & Horsch, 1972
C		430–700[¶]	24–48	"
S	Glyceollin	67	36	Ebel *et al.*, 1976
S	Phaseollin	23**	48	Dixon & Lamb, 1979
S	Phaseollin	50	48	Dixon & Lamb, 1979
S	Phaseollin	14	48	Dixon & Lamb, 1979
S	Phaseollin	34	48	Dixon & Bendall, 1978a
S	Phaseollin	20	48	Dixon & Bendall, 1978a
S	Glyceollin	67	36	Ebel *et al.*, 1976
S	Phaseollin	505	15	Hargreaves & Selby,
	Phaseollidin	31	15	1978
	Phaseollin-isoflavan	207	15	
	Kievitone	151	15	
C	Medicarpin	'Optimum'	48	Gustine *et al.*, 1978
S	Phaseollin	10	48	Dixon & Bendall, 1978a
S	Phaseollin	10	48	Dixon & Bendall, 1978a
S	Phaseollin	58–80	48	Dixon & Bendall, 1978a

194 R.A. DIXON

An absence of expression of race specificity in culture is of less
concern to those interested in phytoalexin biosynthesis. An elicitor
cell culture system which produces phytoalexin at a rate comparable
to that observed in the intact plant has several advantages over the
natural host-parasite system: the lack of chlorophyll and, often,
many of the polyphenols found in the intact plant (Dixon & Bendall,
1978a) allows for easier extraction and assay of biosynthetic enzymes,
(Ebel et al., 1976; Dixon & Bendall, 1978b); the presence of only one
major cell type eliminates the question of enzyme changes occurring
in only a small proportion of the treated cells; and the ease of
application of elicitors and labelled precursors generally allows for
easier interpretation of isotope incorporation and dilution data.
These advantages should allow for rapid progress in the field of
isoflavonoid enzymology, and work on such model systems should provide
information on elicitor modulation of host metabolism, which may then
be re-assessed in the more complex host-parasite system.

REFERENCES

Bailey J.A. (1970) Pisatin production by tissue cultures of *Pisum
    sativum* L. *Journal of General Microbiology* 6, 409-15.
Bailey J.A., Burden R.S. & Vincent G.G. (1975) Capsidiol: an antifungal
    compound produced in *Nicotiana tabacum* and *Nicotiana clevelandii*
    following infection with tobacco necrosis virus. *Phytochemistry* 14,
    597.
Bailey J.A. & Deverall B.J. (1971) Formation and activity of phaseol-
    lin in the interaction between bean hypocotyls (*Phaseolus vulgaris*)
    and physiological races of *Colletotrichum lindemuthianum*.
    Physiological Plant Pathology 1, 435-49.
Davies M.E. (1972) Effects of auxin on polyphenol accumulation and
    the development of phenylalanine ammonia-lyase activity in darkgrown
    suspension cultures of Paul's Scarlet rose. *Planta* 104, 66-77.
Dixon R.A. & Bendall D.S. (1978a) Changes in phenolic compounds
    associated with phaseollin production in cell suspension cultures
    of *Phaseolus vulgaris*. *Physiological Plant Pathology* 13, 283-94.
Dixon R.A. & Bendall D.S. (1978b) Changes in the levels of enzymes
    of phenylpropanoid and flavonoid synthesis during phaseollin
    production in cell suspension cultures of *Phaseolus vulgaris*.
    Physiological Plant Pathology 13, 295-306.
Dixon R.A. & Fuller K.W. (1976) Effects of synthetic auxin levels on
    phaseollin production and phenylalanine ammonia-lyase (PAL) activity
    in tissue cultures of *Phaseolus vulgaris* L. *Physiological Plant
    Pathology* 9, 299-312.
Dixon R.A. & Fuller K.W. (1977) Characterization of components from
    culture filtrates of *Botrytis cinerea* which stimulate phaseollin
    biosynthesis in *Phaseolus vulgaris* cell suspension cultures.
    Physiological Plant Pathology 11, 287-96.
Dixon R.A. & Fuller K.W. (1978) Effects of growth substances on non-
    induced and *Botrytis cinerea* culture filtrate-induced phaseollin
    production in *Phaseolus vulgaris* cell suspension cultures.
    Physiological Plant Pathology 12, 279-88.

Dixon R.A. & Lamb C.J. (1979) Stimulation of *de novo* synthesis of
  L-phenylalanine ammonia-lyase in relation to phytoalexin accumulation
  in *Colletotrichum lindemuthianum* elicitor-treated cell suspension
  cultures of French bean (*Phaseolus vulgaris*). *Biochimica et
  Biophysica Acta* 586, 453-63.
Ebel J., Ayers A.A. & Albersheim P. (1976) Host pathogen interactions
  XII. Response of suspension cultured soyabean cells to the elicitor
  isolated from *Phytophthora megasperma* var. *sojae,* a fungal pathogen
  of soyabean. *Plant Physiology* 57, 775-9.
Glazener J.A. & van Etten H.D. (1978) Phytotoxicity of phaseollin to,
  and alteration of phaseollin by, cell suspension cultures of
  Phaseolus vulgaris. *Phytopathology* 68, 111-7.
Grisebach H. & Hahlbrock K. (1974) Enzymology and regulation of
  flavonoid and lignin biosynthesis in plants and plant cell suspension
  cultures. *Recent Advances in Phytochemistry* 18, 21-52. Academic
  Press, New York.
Gustine D.L., Sherwood R.T. & Vance C.P. (1978) Regulation of phyto-
  alexin synthesis in jackbean callus cultures. Stimulation of
  phenylalanine ammonia-lyase and O-methyltransferase. *Plant
  Physiology* 61, 226-30.
Hargreaves J.A. & Bailey J.A. (1978) Phytoalexin production by
  hypocotyls of *Phaseolus vulgaris* in response to constitutive
  metabolites released by damaged bean cells. *Physiological Plant
  Pathology* 13, 89-100.
Hargreaves J.A. & Selby C. (1978) Phytoalexin formation in cell
  suspensions of *Phaseolus vulgaris* in response to an extract of bean
  hypocotyls. *Phytochemistry* 17, 1099-102.
Helgeson J.P., Budde A.D. & Haberlach G.T. (1978) Capsidiol. A
  phytoalexin produced by tobacco callus tissues. *Plant Physiology*
  61, Supplement p. 53.
Helgeson J.P., Haberlach G.T. & Upper C.D. (1976) A dominant gene
  conferring disease resistance to tobacco plants is expressed in
  tissue cultures. *Phytopathology* 66, 91-6.
Helgeson J.P., Kemp J.D., Haberlach G.T. & Maxwell D.P. (1972) A tissue
  culture system for studying disease resistance: the black shank
  disease in tobacco callus cultures. *Phytopathology* 62, 1439-43.
Ingram D.S. & Robertson N.F. (1965) Interaction between *Phytophthora
  infestans* and tissue cultures of *Solanum tuberosum*. *Journal of
  General Microbiology* 40, 431-7.
Ingram D.S. & Tommerup I.C. (1973) The study of obligate parasites
  in vitro. Fungal Pathogenicity and the Plants' Response (Ed. by
  R.J.W. Byrde & C.V. Cutting), pp. 121-37. Academic Press, London.
Keen N.T. & Horsch R. (1972) Hydroxyphaseollin production by various
  soyabean tissues: a warning against the use of unnatural host-
  parasite systems. *Phytopathology* 62, 439-42.
Mansfield J.W., Hargreaves J.A. & Boyle F.C. (1974) Phytoalexin
  production by live cells in broad bean leaves infected with *Botrytis
  cinerea. Nature* 252, 316-7.
Schenk R.U. & Hildebrandt A.C. (1972) Medium and techniques for
  induction and growth of monocotyledonous and dicotyledonous plant
  cell cultures. *Canadian Journal of Botany* 50, 199-204.

Street H.E. (1973) Gene expression in plant cell cultures. *Biosynthesis and its Control in Plants* (Ed. by B.V. Milborrow), pp. 118-25. Academic Press, London.

Street H.E. (1977) *Plant Tissue and Cell Culture,* 2nd edition. Blackwell Scientific Publications, Oxford.

Warren R.S. & Routley D.G. (1970) The use of tissue culture in the study of single gene resistance of tomato to *Phytophthora infestans*. *Journal of the American Society of Horticultural Science* 95, 266-9.

Zähringer U., Ebel J., Mulheirn L.J., Lyne R.L. & Grisebach H. (1979) Induction of phytoalexin synthesis in soyabean. Dimethylallyl-pyrophosphate:trihydroxypterocarpan dimethylallyl transferase from elicitor-induced cotyledons. *FEBS Letters* 101, 90-2.

The isolation and properties of tomato mesophyll cells and their use in elicitor studies

J.A. CALLOW & J.M. DOW*

Department of Plant Sciences,
University of Leeds, Leeds LS2 9JT

INTRODUCTION

Biochemical studies of the interaction between differentiated plant
tissues and elicitors of disease resistance produced by pathogens are
complicated by a number of problems. It is particularly difficult in
the case of the leaf, with its air-filled spaces and hydrophobic
surfaces, to ensure that elicitor fractions penetrate the tissue and
reach the mesophyll cells. Many of the important recent advances in
understanding cellular recognition processes have been made with
unicellular organisms or with somatic tissue broken down into individual
cells. Therefore, in attempts to simplify his system, the plant
pathologist may adopt a number of alternative approaches. These include
the use of undifferentiated callus, suspension cultures of single cells,
protoplasts, or isolated cells. Each of these approaches has its
advantages and limitations.

   In this paper we show that studies of the interaction between leaf
tissues and fungal elicitors are facilitated by the use of relatively
homogeneous suspensions of isolated mesophyll cells, prepared by a
rapid enzyme maceration technique. Such suspensions can be accurately
pipetted, all cells have equal access to the elicitor and, if required,
the elicitor may be removed by washing. Isolated cells are more robust
and can be prepared more rapidly than protoplasts. Also, they are more
representative of a differentiated tissue than cultured cells derived
from callus.

METHOD OF ISOLATION

The isolation method follows those of Takebe et al. (1968) and Servaites
& Ogren (1977).

---

*Present address: Department of Biology, Box 3AF, New Mexico State
University, Las Cruces, New Mexico 88003.

Ingram D.S. & Helgeson J.P. (1980) *Tissue Culture Methods for Plant Pathologists*

Media

1. Maceration medium: 0.1 M MES-KOH buffer at pH 6.0, containing (g/l): sorbitol, 100; polyvinylpyrrolidone (PVP-40; mol. wt. 40 000; Sigma Chemical Co., London), 20; K_2SO_4, 2.2; Macerase (a crude polygalacturonase preparation from *Rhizopus* spp.; Calbiochem), 10-30.
2. Washing and suspension medium: 0.1 M tricine-HCl buffer at pH 8.4, containing (g/l): KNO_3, 0.5; $Ca(NO_3)_2$, 0.5; $MgCl_2$, 0.2; sorbitol, 54-100.

Procedure

Young leaves of tomato are deveined, surface sterilized in 0.1% sodium hypochlorite containing a wetting agent and washed with sterile distilled water. The leaves are cut into 1 mm wide strips and vacuum-infiltrated with maceration medium for 30 s. Fresh medium is then added and the mixture incubated on a magnetic stirrer for 20 min at 20°C. Undigested pieces of tissue are removed by filtration through muslin and the isolated cells collected on a 40 μm nylon mesh. The cells are then washed and resuspended in medium 2.

INTEGRITY AND PHYSIOLOGICAL ACTIVITY OF CELLS

Preparations contain good yields of separated mesophyll cells, which are bright green in colour. Under phase contrast the cells are highly refractile with a bright, shiny halo (i.e. class I cells). In 0.55 M sorbitol the cells are slightly plasmolysed. A few epidermal cells are also present, but broken cell contents such as chloroplasts are absent.

A number of reactions have been examined as a measure of cell integrity. The isolated cells undergo plasmolysis and deplasmolysis, stain with vital stains such as neutral red, and exclude 2.5% Evans Blue, all of which suggest intact plasma membranes. The rate of CO_2 fixation of these cells (assayed using $NaH^{14}CO_3$, 4180 $\mu E/m^2/s$ irradiance) varies between 12 and 15 μmol CO_2/mg chlorophyll/h (i.e. approximately 20% of *in vivo* rates) when isolated at lower Macerase concentrations. Higher concentrations (30 g/l) produce higher yields of cells but CO_2 fixation rates are much reduced (0.5-2.0 μmol CO_2/mg chlorophyll/h; Dow & Callow, 1979b).

In the experiments described below isolated tomato cells were generally used within a few hours of isolation. At 20°C cells retain integrity for 2-3 days, but then start to lose their bright green appearance. Takebe *et al.* (1968), however, stored tobacco cells for longer periods in the presence of benzyladenine and described the use of cephaloridine and rimocidin as bacteriostats.

EXAMPLES OF THE USE OF ISOLATED TOMATO CELLS IN ELICITOR STUDIES

Recently considerable attention has been paid to the role of secreted

fungal polysaccharides and glycoproteins as elicitors of phytoalexins
and as possible determinants of specificity in host-pathogen interactions. Dow and Callow (1979a) partially characterized a glycopeptide
secreted by the tomato leaf mould pathogen *Fulvia fulva* (syn. *Cladosporium fulvum*). Earlier, crude preparations containing this glycopeptide had been shown to induce a specific pattern of electrolyte leakage
from intact tissue which was closely related to the pattern of resistance and susceptibility in the host-parasite interaction (Van Dijkman
& Kaars Sijpesteijn, 1971, 1973). Dow and Callow (1979b) and Callow
(unpublished data) therefore examined the interaction between the
secreted glycopeptides of *F. fulva* and isolated mesophyll cells derived
from differential tomato hosts, in terms of electrolyte leakage,
phytoalexin elicitation and cell binding. A brief outline of some of
the results of these experiments is reported here to illustrate the use
of isolated cells in this type of study.

Electrolyte leakage (Dow & Callow, 1979b)

Isolated cells of tomato cv. Purdue 135 (resistant to races 0 and 1,2,3)
and cv. Vetomold (resistant to races 0 and 4) were washed and resus-
pended in 0.55 M sorbitol in deionized water, and aliquots containing
60-100 μg chlorophyll (*c*. 15 000 cells) were incubated at 20°C with
desalted culture filtrates or glycopeptide-containing fractions of the
different races (100 μg carbohydrate). The conductance was measured at
intervals, and at the end of the experiment total available ion leakage
was estimated by adding 0.2 ml chloroform and measuring conductance

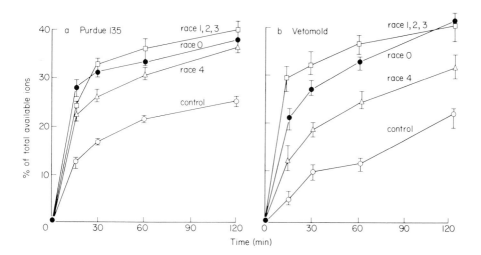

Figure 1. Effect of the high molecular weight fraction of culture
filtrates from different races of *F. fulva* on the leakage of electrolytes from isolated leaf cells of tomato cultivars Purdue 135 (a),
and Vetomold (b), as described by Dow & Callow (1979b). Purdue 135
is susceptible to race 4 and resistant to races 0 and 1,2,3.
Vetomold is susceptible to race 1,2,3 and resistant to races 0 and 4.

after 2 h.   In Fig. 1 values for leakage at different times are
expressed as a percentage of the total available ions.

Fig. 1 shows that high molecular weight culture filtrate fractions of
all three races of *F. fulva* caused a rapid leakage of electrolytes well
above control level.   Purified glycopeptide fractions induced the same
effects.   However, no consistent specific pattern of leakage was
detected between compatible and incompatible combinations, and hence the
suggestions of Van Dijkman and Kaars Sijpesteijn (1971, 1973) cannot be
supported by these experiments.   Degradative treatments showed that the
activity of the glycopeptides in inducing leakage was a function of the
carbohydrate moieties, mannose residues being especially important,
although a role for the protein portion could not be completely ruled
out.

Glycopeptide binding to leaf cells (Dow & Callow, 1979b)

[14]C-labelled glycopeptides were shown to bind reversibly to isolated
tomato leaf cells, but without cultivar or race-specificity (Fig. 2).
The binding could not be ascribed to any single sugar moiety.

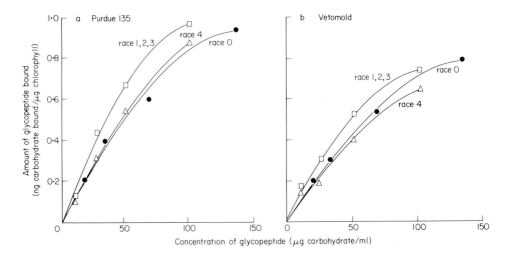

Figure 2.   Binding curves of ^{14}C-labelled glycopeptides from
different races of *F. fulva* to isolated cells of tomato cultivars
Purdue 135 (a) and Vetomold (b).   The glycopeptides were purified
by affinity chromatography as described by Dow & Callow (1979a).
Assays were conducted as described by Dow and Callow (1979b).

Phytoalexin elicitation in isolated cells (Callow, unpublished data)

Isolated cells can be used to test for phytoalexin elicitors.   Fig. 3
shows the results of assays containing 60 000 tomato leaf cells of cv.
Vetomold with various concentrations of glycopeptide elicitor.   After

2 days the suspension medium and 60% methanol extracts of the cells were
partitioned in chloroform. The ability of the two lipophilic fractions
to inhibit germ tube extension of *F. fulva* was then measured in assays
containing 200 spores. Germ tube length was measured after 4 h and
again after a further 12 h. Fig. 3 shows that the glycopeptides were
able to elicit anti-fungal compounds in the isolated cells, some of this
activity being released into the cell medium. However, elicitation was
not race-specific.

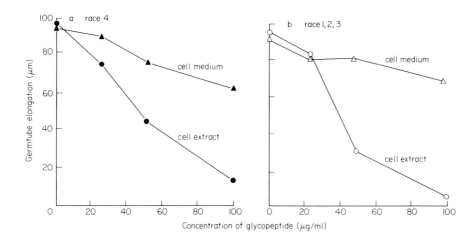

Figure 3. Elicitation of antifungal activity in isolated leaf
mesophyll cells treated with *F. fulva* glycopeptides. Cell suspensions of Vetomold (60 000 cells) were treated with various concentrations of glycopeptide Fraction 1A (Dow & Callow, 1979a) from
culture filtrates of races 4 (a) and 1,2,3 (b). After 2 days, the
cell medium and methanolic extracts of the cells were partitioned in
chloroform. The lipophilic fractions were then added to assays
containing 200 pregerminated spores of race 4 (a) or 1,2,3 (b).
Germ tube length was measured after 4 h, then again after a further
12 h.

REFERENCES

Dow J.M. & Callow J.A. (1979a) Partial characterization of glycopep-
     tides from culture filtrates of *Fulvia fulva* (Cooke) Ciferri (syn.
     Cladosporium fulvum), the tomato leaf mould pathogen. *Journal of
     General Microbiology* 113, 57-66.
Dow J.M. & Callow J.A. (1979b) Leakage of electrolytes from isolated
     leaf mesophyll cells of tomato induced by glycopeptides from culture
     filtrates of *Fulvia fulva* (Cooke) Ciferri, (syn. *Cladosporium fulvum*).
     Physiological Plant Pathology 15, 27-34.
Servaites J.C. & Ogren W.L. (1977) Rapid isolation of mesophyll cells
     from leaves of soyabean for photosynthetic studies. *Plant Physiology*
     59, 587-90.

Takebe I., Otsuki Y. & Aoki S. (1968) Isolation of tobacco mesophyll cells in intact and active state. *Plant and Cell Physiology* 9, 115-24.

Van Dijkman A. & Kaars Sijpesteijn A. (1971) A biochemical mechanism for the gene-for-gene resistance of tomato to *Cladosporium fulvum*. *Netherlands Journal of Plant Pathology* 77, 14-24.

Van Dijkman A. & Kaars Sijpesteijn A. (1973) Leakage of pre-absorbed ^{32}P from tomato leaf discs unfiltered with high molecular weight products of incompatible races of *Cladosporium fulvum*. *Physiological Plant Pathology* 3, 57-67.

The role of tissue culture in the study of crown-gall tumorigenesis

D.N. BUTCHER*, J.L.FIRMIN† & L.M. SEARLE*
*Rothamsted Experimental Station, Harpenden AL5 2JQ
†John Innes Institute, Colney Lane, Norwich NR4 7UH

INTRODUCTION

The effective application of plant tissue culture methods has been largely responsible for the recent advances in the understanding of crown-gall tumorigenesis caused by *Agrobacterium tumefaciens*. The following properties of crown-gall tumours make them ideal experimental material: the tumours may be induced reproducibly on many dicotyledonous species; once induced the tumorous properties are retained in the absence of the bacterium; the tumour tissues can be freed of the bacterium and cultured on simple defined media; and tumour tissues may be grafted onto healthy plants, where they form tumours.

Both tumour and normal cultures may be initiated from the same plant, permitting investigations of the changes that result from tumorigenesis, although the possibility of unpredictable changes occurring during prolonged culture must be considered (Butcher et al., 1975). Early studies showed that tumour tissues, unlike their normal counterparts, grow on media lacking auxins and cytokinins, and it is widely accepted that tumour tissues contain higher concentrations of these compounds. However, these changes in hormone metabolism are not understood.

Experiments with tissue cultures have shown that tumour tissues produce relatively large amounts of unusual amino acid derivatives (opines). The strain of bacterium determines the opines produced: some strains induce tumours containing the octopine family while others induce tumours containing the nopaline family (Fig. 1). A few strains induce tumours which contain neither family. The most recently discovered opine, agropine, frequently accompanies the octopine family, but sometimes occurs on its own. It is produced in much larger amounts than the other opines and is not closely related to them structurally (Fig. 1; Scott et al., 1979; Coxon et al., 1980). In general, strains of *A. tumefaciens* which induce octopine type tumours can utilize the octopine family of opines as a carbon or nitrogen source, while those which induce nopaline type tumours can utilize nopaline and nopalinic acid. Similarly, strains which induce tumours containing agropine can utilize agropine. These observations suggest a transfer

Ingram D.S. & Helgeson J.P. (1980) *Tissue Culture Methods for Plant Pathologists*

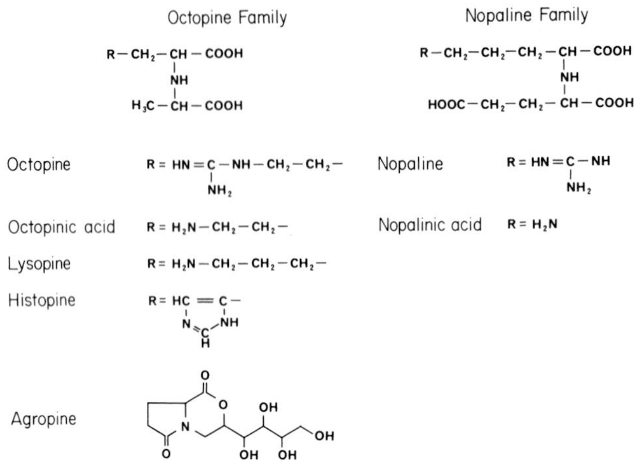

Figure 1. Ti plasmid-determined compounds (opines) present in crown-gall tumour tissues.

of genetic information from the bacterium to the host, and the following evidence (Drummond, 1979) indicates that bacterial plasmid DNA (T-DNA) is transferred to and expressed in the host cell: only oncogenic strains of the bacterium contain the relatively large Ti plasmid ($c.$ 126 x 10^6 mol. wt.); the genes controlling virulence and catabolism of octopine or nopaline as well as the induction of synthesis of these compounds in tumours are all plasmid coded traits; the presence of octopine type T-DNA in a cloned tobacco tumour has been demonstrated by using restriction fragments as probes in a series of hybridization experiments (Chilton *et al.*, 1977).

Some crown-gall tumours of tobacco, known as teratomas, form abnormal shoots as well as completely disorganized tissues. When these shoots are grafted successfully onto decapitated normal plants they are forced into rapid growth and gradually attain a normal morphology. This seems to be a phenotypic reversion of the tumorous state since tissues of the implanted shoot contain T-DNA, synthesize nopaline and become tumorous when returned to tissue culture. However, if the implanted shoot is allowed to flower and set seed, the seeds produce normal plants which lack T-DNA and do not synthesize nopaline. Furthermore, tissue cultures derived from these plants are auxin and cytokinin dependent, a characteristic of normal tissues. These latter observations may indicate a genotypic reversion which appears to occur during meiosis, since haploid tissues derived from anthers have similar properties.

The discovery that T-DNA is transferred from the bacterium to the host cells during tumorigenesis, and that this DNA determines the

synthesis of the opines in transformed tissue, has opened up the
possibility of using the crown-gall system for genetic engineering,
since it may be possible to use the Ti plasmid as a vector for
inserting selected DNA sequences into the plant genome. Work on the
transformation of suspension culture cells and protoplasts with T-DNA
is in progress (Davey *et al.*, this volume).

   This chapter gives instructions for growing bacteria-free crown-gall
tumour tissues in culture from established plants and from aseptically
grown seedlings.

METHODS AND MATERIALS

*Method 1: procedures for obtaining bacteria-free tumour tissue cultures
from established plants (Fig. 2)*

Maintain cultures of *A. tumefaciens* on nutrient agar (17.2 g nutrient
agar, 1 g yeast extract and 5 g sucrose in 1 l water) and store at 4°C.
Transfer the bacteria to liquid medium and incubate at 25°C for 10 h
before using to initiate tumours.

   The age of the host is not critical, but plants with woody stems are
more likely to survive the sterilization treatment necessary to
eliminate the bacterium.

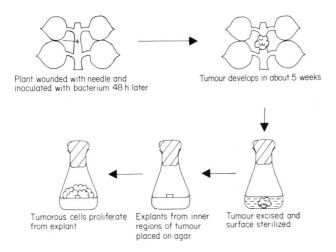

Plant wounded with needle and   Tumour develops in about 5 weeks
inoculated with bacterium 48 h later

Tumorous cells proliferate Explants from inner   Tumour excised and
from explant   regions of tumour   surface sterilized
                            placed on agar

Figure 2.   Initiation of crown-gall tissue cultures from established
plants.

Induction of tumours
   a. Select and surface sterilize the youngest fully elongated inter-
      node with 80% ethanol.

   b. Wound with a sterile needle and cover with moistened cotton wool
      for 48 h.
   c. Uncover the wound, inoculate with a needle previously dipped in
      a bacterial suspension and re-cover for 72 h.
This procedure normally produces small tumours after 2 weeks and these
are left to develop until they are 20-30 mm diam., a suitable size for
the initiation of tissue cultures. At this point it is necessary
to eliminate the bacterium from the tumours and this is more successfully done where the bacterium is confined to the outer regions, as in
tumours of sunflower, artichoke and tobacco.

Initiation of cultures
   a. Surface sterilize tumours by immersing in sodium hypochlorite
      solution (1% available chlorine) or 0.1% mercuric chloride
      solution for 5 min.
   b. Take explants (*c.* 0.5 cm^3) from the inner regions of tumours and
      place on Murashige & Skoog (MS; 1962) medium lacking auxins and
      cytokinins (cf. Wood & Braun, 1961) and solidified with agar (1%).
A small proportion of the explants will give rise to bacteria-free
cultures and the success rate can be increased by adding antibiotics
to the medium (see Method 2). Once established, maintain the cultures
by regular subculturing on media without antibiotics.

*Method 2: procedures for obtaining bacteria-free tumour tissue cultures
from aseptic seedlings (Fig. 3)*

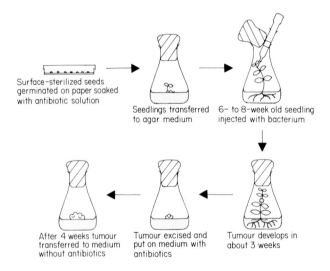

Surface-sterilized seeds
germinated on paper soaked
with antibiotic solution

Seedlings transferred
to agar medium

6- to 8-week old seedling
injected with bacterium

Tumour develops in
about 3 weeks

Tumour excised and
put on medium with
antibiotics

After 4 weeks tumour
transferred to medium
without antibiotics

Figure 3. Initiation of crown-gall tissue cultures from aseptically
grown seedlings.

An alternative method for obtaining tumour cultures from *Nicotiana
tabacum* cv. White Burley is as follows.

a. Surface sterilize seeds by immersing for 5 min in 5% sodium hypochlorite (containing 0.1% Tween 80) and wash three times with sterile distilled water.
b. Put the seeds on sterile filter paper moistened with the antibiotics carbenicillin and vancomycin (200 µg/ml) and leave to germinate in the dark for 7 days at 25°C.
c. Transfer the seedlings aseptically to MS medium solidified with agar (0.6%) and leave to grow in low light (1000 lx) at 20°C for 6-8 weeks.
d. Inject the stems with a suspension of A. tumefaciens and leave for 2-4 weeks, by which time tumours will have developed.
e. Excise the tumours, place on MS medium (containing 100 µg/ml carbenicillin and vancomycin) and leave for 4 weeks.

Thereafter culture the tumour tissues on media without antibiotics and test periodically for bacterial contamination by homogenizing tissue and incubating for 6 days in a complete liquid medium (TY medium of Beringer, 1974).

Grafting experiments with Nicotiana tabacum

Shoots from teratoma cultures of Havana tobacco have been successfully grafted onto rootstocks of Turkish tobacco. The cultures were cloned cell lines derived from tumours induced with the T_{37} strain of A. *tumefaciens* and maintained on MS medium.

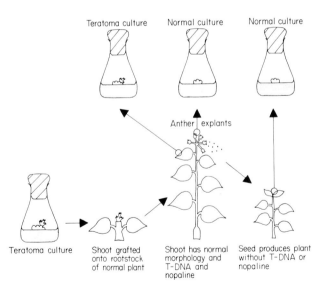

Figure 4. The use of grafting techniques in the demonstration of phenotypic and genotypic reversion of the tumorous state.

Grafting procedure (modified from Braun and Wood, 1976; Fig. 4)
a. Transfer pieces of teratoma culture (*c.* 1 cm^3) to 125 ml Erlenmeyer

flasks containing 15 ml of liquid Mn2 medium (Meins, 1971) and
incubate on a reciprocal shaker in diffuse light at 23-25°C. This
encourages the development of shoots.
b. Isolate shoots and put on basic agar medium (e.g. MS) in diffuse
   light for 1 week.
c. Graft shoots into the cambial regions of cut stems of rootstocks.
A proportion of the implanted shoots will establish a vascular
connection, develop normally and set seed. The use of morphologically
distinct cultivars of tobacco for the teratoma cultures and the stock
plants enables the grafted shoots to be distinguished.

Initiation of tissue cultures from grafted shoots
a. Excise pieces of stem (*c.* 0.5 cm^3) or leaf (*c.* 0.5 cm^2) and wash
   in a 1% solution of Tween 80.
b. Immerse in a 5% solution of calcium hypochlorite for 5 min.
c. Dip in 80% ethanol.
d. Rinse three times in sterilized distilled water.
e. Place on an appropriate agar medium (e.g. MS).

REFERENCES

Beringer J.E. (1974) R Factor transfer in *Rhizobium leguminosarum*.
    Journal of Microbiology 84, 188-98.
Braun A.C. & Wood H.N. (1976) Suppression of the neoplastic state with
    the acquisition of specialized functions in cells, tissues and
    organs of crown gall teratomas of tobacco. *Proceedings of the
    National Academy of Sciences, USA* 67, 1283-7.
Butcher D.N., Sogeke A.K. & Tommerup I.C. (1975) Factors influencing
    changes in ploidy and nuclear DNA levels in cells from normal,
    crown-gall and habituated cultures of *Helianthus annuus* L.
    Protoplasma 86, 295-308.
Chilton M.D., Drummond M.H., Merlo D.J., Sciaky D., Montoya A.L.,
    Gordon M.P. & Nester E.W. (1977) Stable incorporation of plasmid
    DNA into higher plant cells: the molecular basis of crown gall
    tumorigenesis. *Cell* 11, 263-71.
Coxon D.T., Davies A.M.C., Fenwick G.R., Self R., Firmin J.L.,
    Lipkin D. & Janes N.F. (1980) Agropine a new amino acid derivative
    from crown gall tumours. *Tetrahedron Letters* 21, 495-8.
Drummond M. (1979) Crown gall disease. *Nature* 281, 343-7.
Meins Jr.F. (1971) Regulation of phenotypic expression in crown-gall
    teratoma tissues of tobacco. *Developmental Biology* 24, 287-300.
Murashige T. & Skoog F. (1962) A revised medium for rapid growth and
    bioassays with tobacco tissue cultures. *Physiologia Plantarum* 15,
    473-97.
Scott I.M., Firmin J.L., Butcher D.N., Searle L.M., Sogeke A.K.,
    Eagles J., March J.F., Self R. & Fenwick G.R. (1979) Analysis of
    a range of crown gall and normal plant tissues for Ti plasmid-
    determined compounds. *Molecular and General Genetics* 176, 57-65.
Wood H.N. & Braun A.C. (1961) Studies on the regulation of certain
    essential biosynthetic systems in normal and crown-gall tumor cells.
    Proceedings of the National Academy of Sciences, USA 67, 1283-7.

The use of protoplasts for transformation studies with *Agrobacterium tumefaciens* and its plasmids

M.R. DAVEY, J.P. FREEMAN, J. DRAPER &
E.C. COCKING
Department of Botany, University of Nottingham,
University Park, Nottingham NG7 2RD

INTRODUCTION

The Gram-negative bacterium *Agrobacterium tumefaciens* induces crown-gall tumours on dicotyledons and gymnosperms. Transformation of the plant cells to tumorous growth is correlated with the presence, within the inciting bacterium, of large molecular weight $(96-156 \times 10^6)$ tumour-inducing (Ti) plasmids (Van Larabeke *et al.*, 1974; Zaenen *et al.*, 1974; Watson *et al.*, 1975). During transformation the plasmid is transferred, in a way as yet unknown, to recipient cells of the host, where a portion of the plasmid (T-DNA) is integrated into the plant DNA (Chilton *et al.*, 1977; Schell & Van Montagu, 1977) and transcribed (Drummond *et al.*, 1977). The T-DNA codes for tumorigenesis and the synthesis of specific amino-acids (opines, e.g. octopine or nopaline) by tumour cells (Ménagé & Morel, 1964; Petit *et al.*, 1970), the class of opine being determined by the bacterial strain which induced transformation. Tumours excised from host plants can be freed of the inciting bacteria, and proliferate autonomously in tissue culture in the absence of hormones (White & Braun, 1942). Characteristically they multiply to form overgrowths when grafted onto healthy plants (Braun, 1952). Crown-gall disease has been the subject of recent reviews by Lippincott (1977), Drummond (1979) and Gordon (1979).

Current interest in *A. tumefaciens* centres upon the fact that the Ti plasmid is the only known vector which promotes the transfer, integration and expression of foreign DNA in higher plants. It may, therefore, be possible to utilize this plasmid, or smaller T-DNA carrying plasmids which have been constructed by cloning techniques, in the genetic engineering of plant cells. However, before this possibility can be developed it is necessary to demonstrate that plant cells can be transformed reproducibly, preferably in a tissue culture system.

Enzymatically isolated leaf and suspension cell protoplasts provide a convenient single cell system with which to investigate transformation, since they can be maintained in culture (Constabel, 1975; Power & Davey, 1979). Protoplasts of many species undergo division

to form cell colonies from which plants can be regenerated in some
cases. Freshly isolated protoplasts will take up plasmids
(Fernandez et al., 1978; Hughes et al., 1979), with expression in the
case of Ti plasmids in the form of a tumorous response (Davey et al.,
1979, 1980). At a later stage of culture, cells regenerated from
protoplasts can be transformed by intact A. tumefaciens (Davey et al.,
1979; Márton et al., 1979). These recent reports form a basis on which
to extend transformation to other dicotyledons and perhaps also to
monocotyledons. The possibility of being able to transform protoplasts
from monocotyledons is of particular interest, since it may provide
a means of overcoming the barrier to infection by A. tumefaciens
normally encountered in plants such as cereals.

The hormone independence of tumours enables transformed cells to be
selected from a heterogeneous population using hormone-free media.
Characterization of transformation demands an analysis using as many
criteria as possible, none of which are infallible. Simple tests to
verify the crown-gall nature of selected tissues include investigation
of their behaviour on grafting, and analysis to detect the presence of
opines and of the enzymes involved in their synthesis (Otten &
Schilperoort, 1978). This paper summarizes the methods used to transform protoplasts of Petunia spp. by isolated Ti plasmids, and cells
regenerated from protoplasts of Nicotiana spp. by intact A. tumefaciens.

METHODS AND MATERIALS

1. Isolation of Ti plasmids

The following procedure is based on that of Zaenen et al. (1974).
Other methods have been reported (Currier & Nester, 1976).

a. Maintenance of bacterial cultures   Maintain A. tumefaciens
cultures at 28°C in the dark on minimal medium (MM) consisting of
K_2HPO_4 (10.25 g/l), KH_2PO_4 (7.25 g/l), NaCl (0.15 g/l), $MgSO_4.7H_2O$
(0.5 g/l), $CaCl_2.2H_2O$ (0.01 g/l), glucose (5.0 g/l), $FeSO_4.7H_2O$ (2.5
mg/l), Difco-Bacto agar washed twice with distilled water before use
to remove nitrogen (15.0 g/l) and octopine (e.g. strain ACH-5) or
nopaline (T-37) as the nitrogen source (80.0 mg/l). Yeast + beef
extract + peptone medium (YEB) can also be used, consisting of yeast
extract (1.0 g/l), beef extract (5.0 g/l), peptone (5.0 g/l), sucrose
(5.0 g/l) and $MgSO_4.7H_2O$ (0.5 g/l). Subculture the A. tumefaciens
weekly. For long term maintenance grow the bacterium to a high density
in YEB liquid, mix with an equal volume of a sterile 50% v/v solution
of glycerol, and store at -20°C.

b. Preculture for plasmid isolation   Inoculate 120 ml of YEB medium
in a 250 ml flask with one loop of bacteria from MM. Incubate on an
orbital shaker (120 rev/min) overnight at 28°C in the dark. Check the
culture purity by transferring an aliquot to nutrient broth consisting
of Lab Lemco (10.0 g/l), Oxoid peptone (10.0 g/l) and NaCl (5.0 g/l),
at pH 7.0; in this medium contaminants will overgrow A. tumefaciens.

Alternatively plate an aliquot on MM, or use Bernaerts & DeLey's (1963) test for *A. tumefaciens*. Subculture the bacterium daily by transferring one loop of suspension into 120 ml of fresh YEB.

c. Culture for plasmid isolation   Inoculate two 3 l flasks each containing 1.5 l of peptone liquid medium consisting of peptone (4.0 g/l), yeast extract (1.0 g/l) and $MgSO_4.7H_2O$ (0.5 g/l) at pH 7.2 with 60 ml of an overnight preculture. Check that the inoculum does not contain bacterial clumps. Incubate on a reciprocal shaker (120 strokes/min) for 16–18 h at 28°C in the dark until the *D* reaches 1.2–1.5 (Corning Eel Colorimeter, blue filter 621, 455 nm). Harvest the cells by centrifugation for 20 min at 6000 *g*, using an MSE High Speed 18 centrifuge with a 6 x 250 ml rotor. Check the culture purity as in *b*. Wash the pellet in 0.05 M Tris-HCl buffer containing 1.0 M NaCl and store at -20°C until the purity check is complete.

d. Preparation of bacterial lysate   Resuspend 4.5 g of bacteria in 200 ml TE buffer consisting of Tris-HCl (0.05 M) and NaEDTA (0.02 M) at pH 8.0. Add 200 ml of a 3% v/v solution of Sarkosyl NL35 (Ciba-Geigy) in TE buffer. Pre-incubate 300 mg of Protease Type VI (Sigma Chemical Co., London) in 200 ml TE buffer at 37°C for 90 min and add to the mixture. Incubate at 37°C for 75 min until it is clear and homogeneous.

e. Shearing and clearing   Transfer 100 ml portions of lysate to a 400 ml glass beaker and shear the chromosomal DNA, using a Chemap Vibromix (Clandon Scientific Instrument Co.) with a 55 mm diam. plate for 30 s at maximum amplitude. The original lysate is thick and viscous, but the sheared lysate can be poured easily from the beaker. Remove the remaining cells and wall debris by pelleting, using a Beckman Ti 45 (6 x 94 ml) rotor at 25 000 *g* for 30 min. Retain the supernatants.

f. Concentration of bacterial DNA   Layer the supernatants over 1.5 ml glycerol cushions in the Ti 45 rotor. Centrifuge at 27 000 *g* for 5 h or at 17 000 *g* overnight. Remove the top 60 ml of the super-natant with a 60 ml syringe, the next 20 ml with a 20 ml syringe fitted with a silicon rubber tube, and the final volume of supernatant with a Pasteur pipette, without disturbing the 10–15 ml DNA-glycerol layer at the bottom of the tube. Gently resuspend the DNA in TE buffer (6 x 94 ml volumes) and centrifuge onto glycerol in the Ti 45 rotor at 29 000 *g* for 3 h or at 19 000 *g* overnight. Again remove the supernatant and retain the bottom 10 ml.

g. Centrifugation on caesium chloride   To the pooled DNA samples (60 ml) add a solution of 174 g CsCl (Analar; BDH Ltd) in 100 ml TE buffer, followed by a solution of 60 mg of ethidium bromide in 25 ml TE buffer. Stir these in a 1 l beaker in the dark with a 5 cm magnetic bar, at a maximum speed of 50 rev/min, until the mixture is homogeneous. Fill six 38.5 ml polyallomer tubes with the preparation, using a 20 ml syringe fitted with a stainless steel cannula (3.0 mm i.d.). Balance the tubes and centrifuge in a Beckman Ti 70 rotor

at 38 000 g for 48-60 h at 20°C.   Alternatively, if a vertical tube
rotor (VTi 50) is available the centrifugation time can be reduced to
20 h at 38 000 g, and separation improved (use with a slow start and
slow brake).   Examine the tubes under long wave UV light (366 nm); an
upper chromosomal band with a narrow plasmid band beneath should be
visible after centrifugation.   Remove the chromosomal band using a
10 ml syringe and cannula.   Push the latter through the plasmid band
and remove the solution below the plasmid, together with any pellet
(RNA).   Collect each plasmid band in as small a volume as possible
using a clean 5.0 ml syringe.   Pool the plasmid fractions and
re-centrifuge in a Ti 65 (8 x 13.5 ml) angle rotor at 38 000 g for 48
h at 20°C, or in a VTi 65 (8 x 5.3 ml) vertical rotor at 38 000 g for
18 h at 20°C.   Remove the plasmid bands as before, but in 1.0-2.0 ml
final volume.   Retain the plasmid in a sterile 10 ml glass tube with
a ground glass stopper.   All subsequent operations should be carried
out under aseptic conditions.

h. *Removal of ethidium bromide from the plasmid*   Partition the
plasmid preparation against 3 volumes of isopropanol saturated with
CsCl; the ethidium bromide will pass up into the isopropanol layer.
Repeat this operation seven times.   Remove the CsCl by dialysis against
200 ml of 0.05 M Tris-HCl buffer at pH 8.0 or 50 mM sodium citrate
buffer at pH 9.0 (three changes at 4°C, overnight).   The Visking
dialysis tubing 8/32 (Scientific Instrument Centre Ltd, London) should
be soaked in the following solutions before use: 1% w/v Na_2CO_3 (20 min
at 100°C); 10 mM Na_2EDTA (20 min at 22°C); distilled water (two 20 min
changes at 22°C); and sterile 0.1 mM Na_2EDTA (store at 22°C).

i. *Estimation of plasmid yield*   A D of 1.0 at 260 nm is equivalent
to 50 µg of DNA.   Yields of 50-150 µg of plasmid DNA per preparation
can be expected.   Also check the D at 280 nm, since a peak at this
wavelength indicates protein contamination of the DNA.   If contamination
is present remove the protein by phenolization (see *j.* below); if
there is no contamination the plasmid should be stored in a glass
stoppered tube over a few drops of chloroform at 4°C until required.

j. *Phenolization of plasmid DNA*   To 0.5 Kg phenol (stored at -20°C)
add 150 ml of 0.05 M Tris-HCl buffer at pH 8.0 containing 0.2% w/v
8-hydroxyquinoline.   Place in a separating funnel and shake.   Run off
the lower phenol layer.   Place the phenol saturated with buffer beneath
the DNA solution (equal volumes) in a glass stoppered tube.   Shake
gently for 10 min, then spin in a bench centrifuge at 200 g for 5 min
and remove the phenol.   Repeat this process, then extract the DNA
solution by mixing with ether.   Remove the upper ether layer with a
Pasteur pipette, warm the sample to 65°C and very slowly bubble air
through the DNA to drive off any remaining ether.   Finally, redetermine
the spectrum as in *i*.

k. *Sterilization of plasmid DNA by precipitation*   Plasmid prepared
by the procedure given in *a.-i.* is generally sterile.   However, if
any of the preparation steps are performed under non-sterile conditions
the plasmid can be sterilized as follows.   To every 2.0 ml of DNA

solution add 0.2 ml of a sterile 5 M NaCl solution and 3.2 ml of cold
(-20°C) isopropanol. Place at -20°C and leave overnight. Sterilize
a 13.2 ml polyallomer centrifuge tube and cap by immersion in absolute
ethanol for 30 min and then allow them to dry in a laminar flow cabinet
for 30 min. Transfer the plasmid solution to the centrifuge tube, fill
with isopropanol and centrifuge in a Beckman SW 41 (6 x 13.2 ml) rotor
at 10 000 g for 1 h at 4°C. Remove the supernatant and drain the DNA
pellet. Resuspend the DNA to a final concentration of 10 µg in 0.8 ml
TE buffer with 9% w/v mannitol as osmotic stabilizer and leave overnight at 4°C.

1. Visualization of plasmid DNA by electron microscopy   The amount
of supercoiled (intact plasmid DNA) relative to open circular and
linear DNA in a preparation can be determined by electron microscopy.
Prepare the following solutions and mix together in the volumes
indicated: 6 M ammonium acetate (10 µl); 0.5 M Tris-HCl buffer at pH
8.5 (1 µl); 0.05 M Na$_2$EDTA at pH 8.5 (1 µl); Cytochrome C Type V
(Sigma Chemical Co.) dissolved at 2 mg/ml in 0.001 M Na$_2$EDTA at pH 7.2
and millipore filtered before use (8 µl); plasmid DNA adjusted to
0.2-1.5 µg/ml by dilution of an aliquot from *i.* or *k.* (40 µl).
Spread the DNA within 15 min of adding the Cytochrome C. Clean a new
glass slide by thorough washing in soapy water and place in a flat
trough (e.g. the lid of a 9 cm Petri dish) with one end of the slide
resting on one side of the trough. Flood with a 0.25 M solution of
ammonium acetate. Sprinkle a small amount of dry talc onto the surface
of the ammonium acetate just below the point at which the slide enters
the solution, then spread 1 µl of the DNA mixture onto the slide. The
DNA should move down the slide and spread out onto the surface of the
ammonium acetate, pushing the talc before it. The talc enables the
position of the DNA film to be visualized. Allow the preparation to
stand for 20-30 s. Touch the DNA film with a colloidin-coated electron
microscope grid (200 mesh). Stain the DNA by dipping the grid into a
solution of uranyl acetate (5 x 10^{-5} M) in absolute ethanol for 10 s,
dry by touching the edge of the grid to a filter paper, and examine
by dark field transmission electron microscopy. Alternatively,
shadow with platinum at an angle of 6-7° to the horizontal and examine
under bright field transmission microscopy. Supercoiled plasmids
appear as twisted arm-like configurations.

m. Horizontal agarose gel electrophoresis of plasmid DNA   Gel
electrophoresis can be used as an addition or as an alternative to
electron microscopy to check the integrity of plasmid DNA. Prepare
a 0.5% w/v solution of agarose HGT (Miles Laboratories Ltd, Slough)
in electrophoresis buffer consisting of Tris-HCl (89 mM), boric acid
(8.9 mM) and Na$_2$EDTA (2.5 mM) at pH 8.5. Steam for 30 min and maintain
at 60°C for 4 h to stabilize before pouring into the gel apparatus.
Simple inexpensive horizontal slab gel systems have been reported
(McDonell *et al.*, 1977). To 25 µl of plasmid DNA add 5 µl of a
visible dye solution (0.07% w/v bromophenol blue, 7% w/v sodium dodecyl
sulphate, 33% v/v glycerol) and load into the gel using an automatic
pipette of the Finnpipette type. Also load other wells of the gel with
standard concentrations of λ DNA (Boehringer Corporation (London) Ltd,

Lewes, East Sussex). Electrophorese at 60 mA constant current for 3 h
at 4°C. Stain the gel using a 0.01% w/v solution of ethidium bromide
in electrophoresis buffer for 2 h. Examine under long wave UV light
(366 nm). Preparations which are largely supercoiled DNA move only
a short distance into the gel (c. 1 cm) to form a tight band, whereas
contaminating linear DNA is more mobile and forms streaks in the gel.
Comparison of the fluorescence of the plasmid preparation with that of
λ DNA standards also gives an estimate of the concentration of supercoiled DNA in the sample.

2. Transformation of protoplasts by isolated Ti plasmids

The method given here has been used to transform Petunia (P. hybrida x
P. parodii) cell suspension protoplasts with Ti plasmids isolated from
an octopine strain (ACH-5) of A. tumefaciens (Davey et al., 1979, 1980),
but can be adapted for other protoplast systems.

a. Protoplast isolation   General guidelines as to the most suitable
enzyme mixtures for protoplast isolation have been reported (Constabel,
1975; Power & Davey, 1979), but for many tissues the enzyme mixture
must be determined empirically (Hanke, this volume). For Petunia spp.
cells grown in the liquid medium of Uchimiya & Murashige (1974) with
2,4-dichlorophenoxyacetic acid (2,4-D; 2.0 mg/l) and kinetin (0.25
mg/l) are harvested 5 days after transfer. Protoplasts are isolated
using the following mixture: 2% w/v Driselase (Kyowa Hakko Kogyo Ltd,
Tokyo); 2% w/v Meicelase (Meiji Sika Kaisha Ltd, Tokyo); 0.4% w/v
Macerozyme (Yakult Manufacturing Co., Japan); 9% w/v mannitol; the salts
(mg/l) KH_2PO_4 (27.2), KNO_3 (101.0), $CaCl_2.2H_2O$ (1480.0), $MgSO_4.7H_2O$
(246.0), KI (0.16), $CuSO_4.5H_2O$ (0.025); pH adjusted to 5.6. After 16 h
of incubation at 22°C on a horizontal shaker (60 rev/min) pass the
mixture through a 60 μm mesh nylon gauze and pellet the protoplasts by
gentle centrifugation at 90 g for 5 min. Resuspend the pellet in a
solution of the same salts as were used in the enzyme solution, with
the addition of 15% w/v sucrose. Centrifuge at 200 g for 5 min.
Transfer the scum of intact protoplasts to a salts solution with the
addition of 9% w/v mannitol. Count the protoplasts with a haemocytometer and use immediately.

b. Incubation of protoplasts with isolated Ti plasmids   Transfer
4×10^6 protoplasts to a 12 x 100 mm screw-capped tube and centrifuge
at 90 g for 2 min. Remove the supernatant and resuspend the protoplasts
in 1.0 ml TE buffer containing 9% w/v mannitol, 10 μg Ti plasmid and
2 μg poly-L-ornithine HBr (mol. wt. 166 000; Sigma Chemical Co.) for
1 h at 27°C. Poly-L-ornithine stimulates uptake of plasmids into protoplasts (Fernandez et al., 1978). In order to adjust the plasmid
concentration to the value required it may be necessary to concentrate
an aliquot from 1.i. by placing it in a sterile dialysis bag, removing
the excess liquid by evaporation in an air stream, and then resuspending
the plasmid in TE buffer containing osmotic stabilizer. The plasmid
must be aseptic, and before use it is essential that it is checked for
absence of contamination by plating aliquots in MM and nutrient broth.
Set up controls without plasmid. Wash the protoplasts with three 8 ml

changes of TE buffer containing 9% w/v mannitol, plate at 2.0×10^5/ml
in Murashige & Skoog's (MS; 1962) liquid medium with 9% w/v mannitol,
2.0 mg/l 1-naphthaleneacetic acid (NAA) and 0.5 mg/l 6-benzylaminopurine
(BAP), and incubate at 25°C at a light intensity of 700 lx.

c. Selection of transformed protoplasts   Two weeks after uptake
separate the cells by gentle agitation and spread over the surface of
MS medium solidified with 6.0 g/l agar (Sigma Chemical Co.) and
containing reduced hormones (0.44 mg/l NAA; 0.11 mg/l BAP). Maintain
the cells on this medium for 4 months, decreasing the osmotic pressure
during the first 6 weeks by transfer to medium containing 7.0, 4.5, 3.0
and 0.0% w/v mannitol after 1, 2, 4 and 6 weeks respectively.
Subsequently transfer cell colonies to hormone-free medium; those which
continue to proliferate should be maintained on hormone-free medium and
subcultured regularly. Such potential transformants (at a frequency
of *c*. 1 in every 10^5 protoplasts originally plated) can be further
tested for tumorous characteristics. The precise time for which cells
are maintained on reduced hormone media before transfer to media lacking
hormones may vary with the protoplast system, but should be kept to
a minimum to discourage growth of non-transformed tissue by cross-
feeding. Transformation must be weighed against habituation (the
ability of some non-transformed tissues to grow in the absence of
hormones) for any particular experiment. Habituated tissues may appear
on control plates, but these lack other typical crown-gall characteristics such as the ability to form overgrowths when grafted, or to
synthesize opines.

3. Transformation by A. tumefaciens *of cells regenerated from protoplasts*

Cells regenerated from protoplasts isolated from the leaf mesophyll
of *N.cotiana tabacum* cv. *Xanthi* have been transformed by intact *A.
tumefaciens* (Davey *et al.*, 1979). Again, the method given below can
be adapted for use with other protoplast systems.

a. Protoplast isolation, culture and inoculation   To release protoplasts incubate peeled leaf pieces (lower epidermis removed) of tobacco
for 16 h at 25°C in a mixture of 5% w/v Meicelase and 0.5% w/v
Macerozyme, with 13% w/v mannitol and the salts as listed in *2.a.*
Using 9 cm Petri dishes culture the protoplasts in 3 ml of liquid MS
medium over 12 ml of solidified MS medium (0.6% w/v agar), both
supplemented with NAA (2.0 mg/l), BAP (0.5 mg/l) and mannitol (9% w/v).
Adjust the protoplast numbers in the liquid layer to give a plating
density of 2.0×10^5/ml and incubate at 25°C, 700 lx. Transfer cells
regenerated from protoplasts to fresh medium containing 7% w/v mannitol
at day 7, and inoculate with 0.4 ml of a 24 h culture (peptone medium)
of *A. tumefaciens* at day 14. The mixture should contain *c*. 5.0×10^7
bacteria/ml and 0.8×10^2 bacteria/*N. tabacum* cell (the *N. tabacum*
culture usually contains 6.4×10^5 cells/ml by day 14). Incubate the
cultures for 36 h at 25°C in the dark. Set up uninoculated controls.

b. Selection of transformed cells   Transfer the cells to fresh medium (liquid over agar) with 1.0 mg/ml carbenicillin (Pyopen, Beecham Research Laboratories, Brentford) for 4 weeks; during this period reduce the mannitol to 3.5% w/v and finally omit it. Plate the cells in agar-solidified medium containing hormones and antibiotic, spreading the contents of each original dish as four 15 ml aliquots. After 4-6 weeks, using the tip of a scalpel, transfer individual colonies (at least 1000/treatment) to agar-solidified MS medium without hormones, and then subculture every 4 weeks. Those colonies which continue to grow after 3-4 passages on hormone-free medium are potential trans-formants. Test portions of individual colonies at intervals using nutrient broth, and when aseptic omit the antibiotic from the medium. The frequency of transformation is higher when intact *A. tumefaciens* is used than with isolated Ti plasmids, and up to 2% of the colonies transferred to hormone-free medium may be transformed.

4. Simple tests to evaluate the tumorous nature of tissues selected from plasmid-treated protoplasts and A. tumefaciens-*treated cells.*

a. Grafting   Tumours multiply when grafted onto healthy host plants (Braun, 1952). Excise young shoots of glasshouse-grown plants (e.g. *Petunia* spp. or *Nicotiana* spp. depending on the protoplast system used) and surface sterilize with a 10% v/v solution of Domestos bleach (Lever Bros.) for 20 min followed by six washes in sterile water. Cut the shoots into 5 cm lengths and insert a wedge-shaped piece of cultured tissue (40 mg) into a 0.5 cm slit on each stem explant. Cover the graft with a 2.5 cm length of sterile micropore tape (3M Medical Products, Bracknell, Berkshire) which has been sterilized by autoclaving. Insert the stem base into agar-solidified MS medium supplemented with 0.1 mg/l NAA to encourage rooting (40 ml of medium per 6 oz powder-round screw-capped bottle; Beatson Clark and Co., Rotherham, Yorkshire). Incubate at 25°C, 1000 lx. Tissues from plasmid-treated protoplasts, axenic *A. tumefaciens*-treated cells regenerated from protoplasts, and axenic crown-gall tumours (see *4.b.*) proliferate at the graft union and burst through the tape seal. Non-transformed and habituated tissues fail to grow. Excise the overgrowths and return to hormone-free MS medium.

b. Induction of authentic crown-gall tumours on host plants   Surface sterilize stems and insert the bases into medium as in *4.a.* Inoculate them with *A. tumefaciens* by applying a loop of an overnight culture grown in YEB medium to the cut surface of a topped explant, or by pricking with a hypodermic needle previously dipped in the culture. Incubate at 25°C, 1000 lx. Tumours should appear within 7-10 days. Excise the tumours after 3-4 weeks and wash in hormone-free MS medium containing 1.0 mg/l carbenicillin. Transfer to agar-solidified MS medium supplemented with carbenicillin (stock solution prepared, filter sterilized and added to the medium at 40°C). Subculture every 4 weeks. Check sterility by incubating portions of tissue in nutrient broth overnight at 28°C. When the tumours are aseptic maintain them on hormone-free MS medium without antibiotics and use for grafting and opine analysis.

c. Detection of opines in transformed tissues by electrophoresis
Tumours synthesize opines from arginine. Some tumours contain large
amounts of these amino acids, but in others their detection may be more
difficult. Incubation of tumours in an excess of arginine prior to
extraction normally facilitates subsequent analysis. Incubate 200 mg
of tissue in 2 ml of hormone-free MS liquid medium containing 100 mM
L-arginine HCl (BDH Ltd) at 25°C for 48 h, with shaking. Rinse the
tissues with sterile distilled water, blot dry with sterile Kleenex
tissue and homogenize in a micro-centrifuge tube (2 ml capacity) with
a plastic rod (W. Sarstedt Ltd, Leicester). Centrifuge the suspension
at 12 000 g for 2 min, using an Eppendorf or similar instrument.
Retain the supernatant and spot 5 µl samples, 3.5 cm from the anode
and 1.0 cm apart, onto a 12 x 15 cm sheet of Whatman 3MM or MN 214
(Macherey Nagel, 516 Duren, Germany) chromatography paper. Use 4 µl
of a mixture of arginine and octopine or nopaline (depending on the
bacterial strain), each at 0.5 mg/ml, as standard, together with 1 µl
of methyl green in ethanol (migrates just behind the fastest compound,
arginine). Also spot extracts of non-transformed tissue, authentic
crown-gall tumours, and plasmid-treated protoplasts or *A. tumefaciens*-
treated cells mixed with octopine/nopaline standards. Electrophorese
in formic acid (Analar):acetic acid (Analar):water (5:15:80), using
a Shandon flat bed tank for 30 min (360 V, 26 mA at the start; 300 V,
33 mA at the finish), wetting the paper uniformly with cotton wool
soaked in buffer before running. Dry the paper with warm air for 30
min, and develop. Prepare the stain reagent by mixing together,
immediately before use, a solution of 2 mg phenanthrenequinone (BDH Ltd)
in 10 ml absolute ethanol and a solution of 1 g NaOH in 10 ml 60% v/v
ethanol. Allow the anodal end to pass through the reagent before the
arginine spots. Dry with cool air for 10 min and observe under long
wave UV light (366 nm). Octopine and nopaline have a yellowish-green
fluorescence which fades in 15-20 min, although electropherograms can
be stored in airtight plastic bags in the dark at -20°C. Photograph
electropherograms on Ilford Pan F or FP4 black and white film using
a yellow filter.

*d. Detection of lysopine and nopaline dehydrogenase activity in
transformed tissues*   The activity of enzymes involved in the synthesis of opines from arginine in transformed tissues can be determined
using a rapid microscale method as reported by Otten & Schilperoort
(1978).

Acknowledgement

This work was supported in part by grants from the Leverhulme Trust
Fund and from the Science Research Council.

REFERENCES

Bernaerts M.J. & DeLey J. (1963) A biochemical test for crown gall
    bacteria. *Nature* <u>197</u>, 406-7.

Braun A.C. (1952) The crown-gall disease. *Annals of the New York Academy of Sciences* 54, 1153-61.

Chilton M-D., Drummond M.H., Merlo D.J., Sciaky D., Montoya A.L., Gordon M.P. & Nester E.W. (1977) Stable incorporation of plasmid DNA into higher plant cells: the molecular basis of crown gall tumorigenesis. *Cell* 11, 263-71.

Constabel F. (1975) Isolation and culture of plant protoplasts. *Plant Tissue Culture Methods* (Ed. by O.L. Gamborg & L.R. Wetter) pp. 11-6. National Research Council of Canada, Prairie Regional Laboratory, Saskatoon.

Currier T.C. & Nester E.W. (1976) Isolation of covalently closed circular DNA of high molecular weight from bacteria. *Analytical Biochemistry* 76, 431-41.

Davey M.R., Cocking E.C., Freeman J., Draper J., Pearce N., Tudor I., Hernalsteens J.P., De Beuckeleer M., Van Montagu M. & Schell J. (1979) The use of plant protoplasts for transformation by *Agrobacterium* and isolated plasmids. *Proceedings of the 5th International Protoplast Symposium 1979,* pp. 425-30. Akadémiai Kiadó, Budapest.

Davey M.R., Cocking E.C., Freeman J., Pearce N. & Tudor I. (1980) Transformation of *Petunia* protoplasts by isolated *Agrobacterium* plasmids. *Plant Science Letters* 18, 307-13.

Drummond M.H. (1979) Crown gall disease. *Nature* 281, 343-7.

Drummond M.H., Gordon M.P., Nester E.W. & Chilton M-D. (1977) Foreign DNA of bacterial plasmid origin is transcribed in crown gall tumours. *Nature* 269, 535-6.

Fernandez S.M., Lurquin P.F. & Kado C.I. (1978) Incorporation and maintenance of recombinant-DNA plasmid vehicles pBR 313 and pCR1 in plant protoplasts. *Federation of European Biochemical Societies Letters* 87, 277-82.

Gordon M.P. (1979) Plant Tumours. *Biochemistry of Plants Vol. 6. Proteins and Nucleic Acids* (Ed. by A. Marcus). Academic Press, New York and London (in press).

Hughes B.G., White F.G. & Smith M.A. (1979) Fate of bacterial plasmid DNA during uptake by barley and tobacco protoplasts. II. Protection by poly-L-ornithine. *Plant Science Letters* 14, 303-10.

Lippincott J.A. (1977) Molecular basis of plant tumour induction. *Nature* 269, 465-6.

McDonell M.W., Simon M.N. & Studier F.W. (1977) Analysis of restriction fragments of T7 DNA and determination of molecular weights by electrophoresis in neutral and alkaline gèls. *Journal of Molecular Biology* 110, 119-46.

Márton L., Wullems G.J., Molendijk L. & Schilperoort R.A. (1979) *In vitro* transformation of cultured cells from *Nicotiana tabacum* by *Agrobacterium tumefaciens. Nature* 277, 129-31.

Ménagé A. & Morel M.G. (1964) Sur la présence d'octopine dans les tissus de crown-gall *Comptes rendus hebdomadaire des séances de l'Académie des sciences* 259, 4795-6.

Murashige T. & Skoog F. (1962) A revised medium for rapid growth and bioassays with tobacco tissue cultures. *Physiologia Plantarum* 15, 473-97.

Otten L.A.B.M. & Schilperoort R.A. (1978) A rapid microscale method
    for the detection of lysopine and nopaline dehydrogenase activities.
    Biochimica et Biophysica Acta <u>527</u>, 497-500.
Petit A., Delhaye S., Tempé J. & Morel M.G. (1970) Recherches sur les
    guanidines des tissus de crown gall. Mise en évidence d'une relation
    biochimique spécifique entre les souches d'*Agrobacterium tumefaciens*
    et les tumeurs qu'elles induisent. *Physiologie Végetale* <u>8</u>, 205-13.
Power J.B. & Davey M.R. (1979) *Laboratory Manual: Plant Protoplasts
    (Isolation, Fusion, Culture, Genetic Transformation)*. Department
    of Botany, University of Nottingham.
Schell J. & Van Montagu M. (1977) Transfer, maintenance and expression
    of bacterial plasmid Ti DNA in plant cells transformed with
    Agrobacterium tumefaciens. *Genetic Interaction and Gene Transfer.
    Brookhaven Symposium in Biology* <u>29</u>, pp. 36-49.
Uchimiya M. & Murashige T. (1974) Evaluation of parameters in the
    isolation of viable protoplasts from cultured tobacco cells. *Plant
    Physiology* <u>54</u>, 936-44.
Van Larabeke N., Engler G., Holsters M., Van Den Elsaker S., Zaenen I.,
    Schilperoort R.A. & Schell J. (1974) Large plasmid in *Agrobacterium
    tumefaciens* essential for crown-gall inducing ability. *Nature* <u>252</u>,
    169-70.
Watson B., Currier T.C., Gordon M.P., Chilton M-D & Nester E.W. (1975)
    Plasmid required for virulence of *Agrobacterium tumefaciens*.
    Journal of Bacteriology <u>123</u>, 255-64.
White P.R. & Braun A.C. (1942) A cancerous neoplasm of plants:
    autonomous bacteria-free crown-gall tissue. *Cancer Research* <u>2</u>, 597.
Zaenen I., Van Larebeke N., Teuchy H., Van Montagu M. & Schell J. (1974)
    Supercoiled circular DNA in crown-gall inducing *Agrobacterium*
    strains. *Journal of Molecular Biology* <u>86</u>, 109-27.

Section 5
Methods for the production of novel disease-resistant plants

Tissue culture methods in plant breeding

P.R. DAY

Plant Breeding Institute,
Cambridge CB2 2LQ

INTRODUCTION

This paper presents some selected examples to illustrate the application
of tissue culture methods in plant breeding. A variety of reviews on
this and related subjects have appeared during the last 7 years. Some
useful edited collections of papers deal with the culture of cells,
tissues and organs (Street, 1974; Ledoux, 1975; Barz et al., 1977;
Reinert & Bajaj, 1977), haploids (Kasha, 1974), protoplasts (Peberdy et
al., 1976) and cell organelles (Birky et al., 1975; Bücher et al.,
1976). Other relevant reviews include those of Nickell & Heinz (1973),
Day (1977) and Thomas et al. (1979) of a general nature, and that of
Brettell & Ingram (1979) concerning the production of disease-resistant
crops. Although it does not deal with tissue cultured aseptically, the
review by Broertjes & Van Harten (1978) on mutation breeding of
vegetatively propagated crops is also of interest.

MULTIPLICATION

The majority of agronomic crop plants are propagated by seed but an
important minority (potato, sweet potato, sugar cane etc.) are vegetatively propagated. For ornamental crops vegetative propagation is more
important, largely because the value of individual plants allows a much
greater investment in the labour and facilities required. Some
agronomic and horticultural crop plants either have no natural method
of vegetative multiplication (e.g. oil palm) or can only be multiplied
at very slow rates (e.g. orchids). The use of tissue culture methods
has made an impressive contribution to the multiplication of such
plants, enabling the production of what Murashige (1977) has described
as "clonal crops".

The first steps in this direction were taken in the early 1950s,
when Morel demonstrated the use of shoot-tip cultures to produce virus-
free plants of dahlia, carnation and potato. Applying this same method
to produce virus-free clones of Cymbidium by culturing axillary buds,
Morel established the method of meristem culture which revolutionized

Ingram D.S. & Helgeson J.P. (1980) *Tissue Culture Methods for Plant Pathologists*

the propagation of orchids (Morel, 1964). There is now an industry
based on this method of propagation; its impact on the improvement of
a horticultural crop is well illustrated by the work of Murashige on
the propagation of Gerbera (Murashige et al., 1974). Traditionally
grown from seed, this ornamental produces mixed colours. Controlled
hand pollinations yield one seed per floret, an inherent limitation in
breeding members of the Compositae. The technique of meristem culture
provides an alternative route. Plants are allowed to flower and single
plants with the desired colours selected. The apical meristem of a
lateral branch still showing vegetative growth is aseptically dissected
out and, when incubated on a suitable medium in a culture tube under
lights, grows and forms a cluster of shoots. These can be dissected
apart and the process repeated many times. When plants are required
the small shoots are transferred to a medium which induces rooting,
and sturdy plantlets form which will grow in soil and eventually produce
flowers all of the selected colour of the parent plant. More than 10^6
rooted plants can be formed from a single plant in a year. Selection
and clonal propagation of desired individual plants provides a more
rapid means of achieving what a plant breeder might do although
probably not as economically in terms of man-hours and material
resources. Since Californians want to buy Gerbera plants to put in
their borders, and not seed to sow and transplant themselves, this
method of propagation fits in well with horticultural trade practice
for bedding plant production.

For oil palm, tissue culture has provided the possibility of clonal
propagation. Since the palm has only a single meristem that does not
divide, meristem culture cannot be used. Instead callus cultures are
prepared from seedling leaf base or root tissue and are induced to form
adventitious shoots. These form roots and can be transplanted to give
rise to clones of identical plants. The clones can be used in spacing
and yield trials since they provide for the first time the opportunity
for replication in the trials designs (Corley et al., 1977).

For the time being tissue culture propagation has a limited use
in agronomic crops, although its application to potato has greatly
simplified the preservation of genetic stocks in germplasm banks.
Individual clones may be maintained as meristem cultures free from
the rigours and selection pressure of field plots. The method also has
the potential of increasing the rate of multiplication of new cultivars
selected originally from seedlings. At the Plant Breeding Institute
meristem culture could provide a means of multiplication and stock
maintenance that would greatly lessen present dependence on tuber
multiplication in field plots in the virus-free protected area of
Scotland. The protected area of the culture vessel might well become
important by offering economies in scale, time and travel.

Tissue culture propagation methods have yet to make much impact on
crops that are most economically cultivated from true seed. However,
if and when the problem of inducing embryogenesis in cell suspension
cultures can be solved for plants other than wild carrot, there is the
possibility of producing artificial seeds. These would be embryoids,

pelleted in a nutrient and protective coat, that can be sown direct
in the soil and which 'germinate' there (Murashige, personal communication). Such a technique would be extremely valuable for a biennial
crop plant such as sugar beet, where breeders have to maintain
cytoplasmically male sterile lines, or fully fertile non-restorer lines
used to maintain and improve the male sterile lines, or diploid or
tetraploid pollinators for hybrid seed production. In addition there
must also be selection for monogerm seed, freedom from bolting,
resistance to disease and insects and, most important of all, high
yield and high sugar content; a complex breeding task that could be
simplified by the successful clonal propagation of a superior phenotype,
avoiding all the complexities of genetic maintenance and seed transmission. The problem of selection would remain, but the risk of
genetic uniformity from clonal monocultures could be relieved by
judiciously mixing genetically different components to produce multi-
clones.

Since cereal breeding systems are simpler than those of crops such
as sugar beet it is easier to fix the complement of characters required
by the breeder, and even if artificial seeds could be produced it is
doubtful whether they would supplant true seeds. The most significant
impact of tissue culture on cereal breeding is probably through the use
of embryo culture to recover haploid barley (Kasha, 1974). When
Hordeum vulgare is crossed as female with *H. bulbosum*, fertilization
occurs readily, but *H. bulbosum* chromosomes are eliminated from the
embryo by a mechanism that is still not understood. At the same time
the endosperm fails to develop after 2 or 3 days. If the small slow-
growing embryos are rescued to a suitable culture medium, a high
proportion develop into haploid plants which can be doubled by
colchicine treatment. Seed produced by such doubled monoploids
produces uniform populations confirming its homozygosity. When F_2
plants from crosses between different *H. vulgare* lines are used as
female parents, the resultant doubled monoploids thus provide a series
of different homozygous lines among which the breeder may be able to
select a finished variety without further inbreeding, rogueing or
purification.

In other genera haploids can be produced by anther culture. In
wheat this is possible with some difficulty, but in the family
Solanaceae it occurs with much greater readiness (Sunderland, this
volume). In some crop species haploids after chromosome doubling are
useful for producing inbred lines for F_1 hybrid production.

VARIATION

Although producing enough of a desirable and uniform product is often
an important plant breeding objective, an equally, if not more important, objective is to introduce variation to select an improvement,
sometimes in one but generally in several characters. Some ways of
achieving this through tissue culture are described below.

Embryo culture

Many difficult or sterile crosses, like the *H. vulgare* x *H. bulbosum* cross, result from a failure of endosperm development in the fertilized ovule. As a consequence embryo growth is arrested and no viable seeds are formed. Excision of the immature embryo and culturing it on an artificial medium which substitutes for the endosperm will often suffice to produce a viable hybrid plant. By this means a wide variety of new characters have been introduced through hybridization. The method has been particularly useful in introducing genes for disease resistance. An early example was the introduction of resistance to *Fulvia fulva* from *Lycopersicon peruvianum* into cultivated tomato (*Lycopersicon esculentum*; Alexander *et al.*, 1942).

Protoplasts

A number of authors have remarked on the degree of variation observed among tissue cultures derived from a common source and among plants regenerated from callus. In sugar cane, a clonally propagated crop, this variation among clones derived from single cells has been put to some practical use and has resulted in selection of mosaic virus resistant clones (Nickell & Heinz, 1973) and clones resistant to eyespot (*Drechslera sacchari*), Fiji disease (probably a virus) and downy mildew (*Sclerospora sacchari*) (Heinz *et al.*, 1977). Treatments with mutagens did not enhance the rate of recovery of resistant clones. Cytological examination revealed a considerable variation in chromosome number in the form of mosaics in callus cultures and also in the clones them- selves. Members of the genus Saccharum are polyploids (x=10) with chromosome numbers ranging from 2n = 40 to 2n = *c*. 200. Much of the variation observed among the clones derived from tissue cultures has been attributed to chromosomal reassortment resulting from isolation of components of the cytological mosaics. However, some cane clones give few or no variants when cultured and in any case most cane clones are remarkably stable when propagated vegetatively by organized meristems.

Potato clones are similarly regarded as stable, but seed producers in Scotland help ensure this stability by rogueing all offtype plants in fields certified for seed tuber production. Little or no attention has been paid to the occasional variants that appear since there has been little likelihood that they would have any economic significance. Some recent work by Shepard may change this point of view.

In the course of studies with potato viruses, Shepard began examin- ing the variation among potato clones regenerated from protoplasts prepared from leaf mesophyll tissue of the variety Russet Burbank. Released in 1873 in Massachusetts, the variety Burbank was one of 23 seedlings produced by a variety named Early Rose. The selection was made by Luther Burbank, and the mutation that gave rise to its russet, dark-skinned tubers occurred around 1910. Russet Burbank currently accounts for about 39% of the North American potato crop. Favoured by processors and ware producers alike, it has dominated the market in

spite of a number of defects that include susceptibility to potato
leaf roll virus, spindle tuber viroid, bacterial ring rot and blight
(*Phytophthora infestans*). Russet Burbank produces misshapen tubers
in heavy soils, and the haulms are somewhat indeterminate in growth
habit. Its geographical distribution is sharply limited by high summer
temperatures at the southern end of its range in the western and central
potato growing area of the US.

By carefully refining their techniques, Shepard and his colleagues
were able to define the conditions for very high rates of plant
regeneration from isolated protoplasts (Shepard & Totten, 1977). The
results of their work are of considerable interest and are reported
in two papers (Shepard, 1980; Shepard *et al.*, 1980).

A variable percentage of the plants derived from protoplasts were
clearly aberrant. The aberrant forms included dwarfs, plants with
variations in chlorophyll intensity and in the number, form and size
of leaves, and plants with aerial tubers. Many aberrants were recognized during shoot regeneration and discarded. Cytological examination
showed that many had abnormal chromosome numbers. However, another
class of variants proved to be of more interest.

In 1977, 1700 clones derived from protoplasts (protoclones) were
planted in Montana. After field examination and harvest 396 were
selected, grown at two sites in Idaho and by further selection reduced
to 60 clones that were evaluated in 1979 in Florida as well as at the
two sites in Idaho. Many of these were also tested at two other sites
(five in all).

Examples of variants that were stable over three tuber generations
include the following.

1. Habit   Plants had a determinate, more compact habit and shorter
internodes.

2. Tuber characteristics   There were high and low frequencies of
misshapen tubers; three clones produced twice as many tubers as the
parent, and very few were misshapen. Preliminary data indicate that
these clones may outyield the parent clone by as much as 80% but further
trials are needed to confirm this. Two clones produced smooth, white-
skinned tubers. One clone formed tubers with more elongate profiles
than the parent.

3. Maturity date   Russet Burbank is a late maturing variety. Some
protoclones were earlier, others were up to 4 weeks later.

4. Photoperiod requirements   In a controlled environment, plants of
Russet Burbank flower in long days (16 h) and most flower buds fall off
before they complete development. The berries also fall off, so that
the variety is very difficult to use as a female parent in breeding.
Some protoclones produced many berries but these clones were as self-
sterile as the parent variety.

5. Disease resistance - early blight   Matern *et al.* (1978) tested
leaves of 500 protoclones for resistance to *Alternaria solani* by
spraying a suspension of spores and mycelium on the lower leaves of
field-grown plants and enclosing them in plastic bags. Five resistant
clones were recovered, three with intermediate and two with full resistance. Leaves from these clones were tested against a toxin isolated
from *A. solani* culture filtrate. The cut petioles were placed in the
toxin solution which was imbibed as a result of transpiration.
Comparable reactions to those obtained by direct inoculation were
claimed to have been observed.

6. Disease resistance - late blight   Rooted vegetative cuttings were
prepared from 800 protoclones and, when 6-8 cm high, were inoculated
with race 0 of *P. infestans*. After 14 days 20 plants (*c.* 2%) were not
killed. Their symptoms ranged from near death (they showed regrowth
from a basal bud) to scattered necrotic lesions. Two further tests
with additional rooted cuttings gave the same result. Tests of eight
clones with race 1,2,3,4 gave similar results (one clone resistant to
race 1 was susceptible to race 1,2,3,4; one clone with low resistance
to race 1 was much more resistant to race 1,2,3,4). Tubers collected
from uninoculated plants of the resistant clones showed a similar
reaction in the next generation. The resistance was described as more
closely resembling minor gene resistance. The necrotic lesions
increased in size and supported some sporulation. None of the clones
resistant to *A. solani* were resistant to blight. The reciprocal test
was not carried out.

   Shepard wisely advocates caution in interpreting these preliminary
results. Clones improved in one respect may be deficient in another,
and this possibility can only be eliminated by full testing and proper
performance evaluation. Protoplasts were successfully cultured and
raised to calluses of six other potato varieties, but there is no
published data on protoclones from these. Behnke (1979), working with
callus from young leaves of dihaploid (2n = 24) potato, recovered five
calluses resistant to media that incorporated toxic culture filtrates
of *P. infestans*. Wenzel *et al.* (1979) examined protoclones from two
dihaploids and also obtained some variation in morphology but less than
that claimed by Shepard and his colleagues.

   Clearly if Shepard's findings with the potato variety Russet Burbank
have a general validity, there is the possibility of exploiting this
variation to secure improvements in already well adapted UK varieties.
At the present time the nature of protoclone variation is not
completely understood, but it seems likely that it is the result of
chromosomal rearrangements (Shepard, personal communication). Whether
these are induced by the operations of producing and culturing protoplasts or whether they are a feature of potato mesophyll cells is not
known.

Protoplast fusion

The first somatic hybrid produced by protoplast fusion duplicated the known sexual hybrid between *Nicotiana glauca* and *Nicotiana langsdorfii* (Carlson *et al.*, 1972). The selection of hybrid protoplasts and cells was based on their lack of dependence on an exogenous supply of auxin in the culture medium. Both the parent species require auxin for growth in culture. Power *et al.* (1980) have succeeded in obtaining a somatic hybrid between *Petunia parviflora* (2n = 18) and *Petunia parodii* (2n = 14) which cannot be produced sexually. Selection was based on the fact that *P. parodii* protoplasts do not develop calluses while those of *P. parviflora* do. Irradiation of seed of *P. parviflora* enabled the recovery of a nuclear recessive albino mutant from which callus and hence protoplasts could be obtained. Somatic hybrids thus appear as green calluses among a background of albino cells and calluses. The hybrid nature of regenerated flowering plants was confirmed by observing the somatic chromosome number (31) which approached the expected 4n = 32 of the amphidiploid and by the presence of Fraction 1 protein subunits characteristic of both parents.

A tomato x potato hybrid produced by Melchers *et al.* (1978) is another, earlier, example suggesting that protoplast fusion may soon extend the limits placed by sexual compatibility and sterility on interspecific hybridization.

TRANSFORMATION

The discovery that crown-gall disease is caused by the transfer to host cells of part of a large plasmid carried by oncogenic (tumour forming) strains of the bacterium *Agrobacterium tumefaciens* suggests that this natural method of transformation may be adapted for genetic manipulation. One of the most recent of many reviews is that by Drummond (1979). Tumour tissue is distinguished by its ability to grow in culture without added auxins and cytokinins and by its ability to synthesize one or more of several derivatives of basic amino acids, called opines, that are not found in healthy tissue (Butcher *et al.*, this volume). Tumour tissue contains several copies of DNA of about 12 kilobases in size (T-DNA) identical to a small part of the tumour inducing (Ti) plasmid. Opine production by host tissue favours growth of the *A. tumefaciens*, which can use them as carbon and nitrogen sources, to the exclusion of other micro-organisms which cannot. In nature crown-gall is limited to dicotyledonous hosts. Recent experiments have shown that plant protoplasts can be transformed with plasmid DNA from *A. tumefaciens* (Davey *et al.*, this volume). Some tumours are teratomas rather than unorganized tissue. The Ti plasmid controlling synthesis of the opine nopaline produces teratomas. Tobacco teratomas can give rise to flowering shoots which continue to carry T DNA, synthesize nopaline and revert to tumour tissue when cultured. However, following meiosis, these characters appear to be lost and are not found in seedlings or in haploid plants derived from pollen. Unless integration of T and associated DNA can occur which survives gamete formation, it seems likely that this form of transformation will be restricted to vegetatively propagated plants.

There seems little doubt that the rapid progress in plant molecular biology made possible by the application of recombinant DNA techniques will soon lead to other methods of transformation to secure directed changes in plants. Although for many crops there is at present no shortage of genetic variation from related species the transfer of useful information is often slow and difficult and complicated by the accompanying useless, sometimes detrimental, genes transferred at the same time. The challenge of the 1980s is clearly to better understand the exact nature of the changes we wish to make in our crops so that when we have the tools we will know how to use them.

REFERENCES

Alexander L.J., Lincoln R.E. & Wright V. (1942) A survey of the genus *Lycopersicon* for resistance to the important tomato diseases occurring in Ohio and Indiana. *Plant Disease Reporter, Supplement* 136, 49-85.
Barz W., Reinhard E. & Zenk M.H. (1977) *Plant Tissue Culture and its Biotechnological Application.* Springer-Verlag, Berlin.
Behnke M. (1979) Selection of potato callus for resistance to culture filtrates of *Phytophthora infestans* and regeneration of resistant plants. *Theoretical and Applied Genetics* 55, 69-71.
Birky Jr. C.W., Perlman P.S. & Byers T.J. (1975) *Genetics and Biogenesis of Mitochondria and Chloroplasts.* Ohio State University Press, Columbus.
Brettell R.I.S. & Ingram D.S. (1979) Tissue culture in the production of novel disease-resistant crop plants. *Biological Reviews* 54, 329-45.
Broertjes C. & Van Harten A.M. (1978) *Application of Mutation Breeding Methods in the Improvement of Vegetatively Propagated Crops.* Elsevier, Amsterdam.
Bücher T., Neupert W., Sebald W. & Werner S. (1976) *Genetics and Biogenesis of Chloroplasts and Mitochondria.* North Holland, New York.
Carlson P.S., Smith H.H. & Dearing R.D. (1972) Parasexual interspecific plant hybridisation. *Proceedings of the National Academy of Sciences, USA* 69, 2292-4.
Corley R.H.V., Barrett J.N. & Jones L.H. (1977) Vegetative propagation of oil palms via tissue culture. *Oil Palm News* 22, 2-7.
Day P.R. (1977) Plant genetics: increasing crop yield. *Science* 197, 1334-9.
Drummond M. (1979) Crown gall disease. *Nature* 281, 343-7.
Heinz D.J., Krishnamurthi M., Nickell L.G. & Maretzki A. (1977) Cell tissue and organ culture in sugarcane improvement. *Applied and Fundamental Aspects of Plant Cell, Tissue and Organ Culture* (Ed. by J. Reinert & Y.P.S. Bajaj), pp. 1-17. Springer-Verlag, Berlin.
Kasha K.J. (1974) *Haploids in Higher Plants, Advances and Potential.* University Press, Guelph, Ontario.
Ledoux L. (1975) *Genetic Manipulations with Plant Material.* Plenum, New York.

31

Matern U., Strobel G. & Shepard J. (1978) Reaction to phytotoxin in a potato population derived from mesophyll protoplasts. *Proceedings of the National Academy of Sciences, USA* 75, 4935-9.

Melchers G., Sacristan M.D. & Holder A.A. (1978) Somatic hybrid plants of potato and tomato regenerated from fused protoplasts. *Carlsberg Research Communications* 43, 203-18.

Morel G. (1964) Tissue culture - A new means of clonal propagation in orchids. *American Orchid Society Bulletin* 33, 473-8.

Murashige T. (1977) Clonal crops through tissue culture. *Plant Tissue Culture and its Biotechnological Application* (Ed. by W. Barz, E. Reinhard & M.H. Zenk), pp. 392-403. Springer-Verlag, Berlin.

Murashige T., Serpa M. & Jones J. (1974) Clonal multiplication of Gerbera through tissue culture. *Horticultural Science* 9, 175-80.

Nickell L.G. & Heinz D.J. (1973) Potential of cell and tissue culture techniques as aids in economic plant improvement. *Genes, Enzymes and Populations* (Ed. by A.M. Srb), pp. 109-28. Plenum, New York.

Peberdy J.F., Rose A.H., Rogers H.J. & Cocking E.C. (1976) *Microbial and Plant Protoplasts*. Academic Press, New York.

Power J.B., Berry S.F., Chapman J.V. & Cocking E.C. (1980) Somatic hybridisation of sexually incompatible Petunias: *Petunia parodii, Petunia parviflora*. *Theoretical and Applied Genetics* 57, 1-4.

Reinert J. & Bajaj Y.P.S. (1977) *Applied and Fundamental Aspects of Plant Cell, Tissue and Organ Culture*. Springer-Verlag, Berlin.

Shepard J.F. (1980) Mutant selection and plant regeneration from potato mesophyll protoplasts. *Emergent Techniques for Genetic Improvement of Crops* (Ed. by I. Rubenstein, B. Gengenbach & C.E. Green) University of Minnesota Press, Minneapolis (in press).

Shepard J.F., Bidney D. & Shakim E. (1980) Potato protoplasts in crop improvement. *Science* 208, 17-24.

Shepard J.F. & Totten R.E. (1977) Mesophyll cell protoplasts of potato. Isolation, proliferation and plant regeneration. *Plant Physiology* 60, 313-6.

Street H.E. (1974) *Tissue Culture and Plant Science*. Academic Press, New York.

Thomas E., King P.J. & Potrykus I. (1979) Improvement of crop plants via single cells *in vitro* - an assessment. *Zeitschrift für Pflanzenzüchtung* 82, 1-30.

Wenzel G., Schieder O., Przewozny T., Sopory S.K. & Melchers G. (1979) Comparison of single cell culture derived *Solanum tuberosum* L. plants and a model for their application in breeding programs. *Theoretical and Applied Genetics* 55, 49-55.

Selection of maize tissue cultures resistant to *Drechslera (Helminthosporium) maydis* T-toxin

R. BRETTELL*, D.S. INGRAM† & E. THOMAS‡

*Friedrich Miescher-Institut, PO Box 273
CH-4002 Basel, Switzerland
†Botany School, Downing Street, Cambridge CB2 3EA
‡Rothamsted Experimental Station, Harpenden,
Hertfordshire AL5 2JQ

INTRODUCTION

There are several ways in which the selection of mutants may be achieved using plant cell cultures. These are discussed by Strauss *et al.* (this volume) and include the possibility of obtaining toxin-resistant clones *in vitro*, leading to the regeneration of toxin-resistant plants. These techniques might find useful application in the selection of plants resistant to microbial toxins, particularly where they are a determinant of pathogenicity. This is certainly the case in a small number of plant diseases in which the pathogen produces a host-specific toxin. Examples include victoria blight of oats caused by *Drechslera (Helminthosporium) victoriae* (Pringle & Scheffer, 1964), early blight of potato caused by *Alternaria solani* (Matern *et al.*, 1978) and southern corn leaf blight of maize caused by *Drechslera (Helminthosporium) maydis* race T (Smedegard-Petersen & Nelson, 1969). Here failure by the pathogen to produce toxin results in a loss of virulence, and it is likely that insensitivity of the host to the toxin would result in disease resistance. Additionally, in other diseases such as the wildfire disease of tobacco caused by *Pseudomonas tabaci* (Patil, 1974), where a non-specific toxin is produced by the pathogen, toxin resistance in the host plant might be correlated with a degree of resistance to the disease.

A single example of how toxin resistance may be selected using a culture derived from a susceptible genotype of the host plant is described here with maize and the disease southern corn leaf blight (SCLB), using the technique of recurrent selection without a prior mutagenic treatment (Maliga *et al.*, 1973; Gengenbach & Green, 1975). The virulent form of the disease is caused by race T of *D. maydis*, which produces T-toxin in culture and in the leaves of infected plants (Smedegard-Petersen & Nelson, 1969; Lim & Hooker, 1972). When applied to leaf wounds or injected, T-toxin is found to induce characteristic symptoms in plants with Texas male-sterile (Tms) cytoplasm, but not in those with normal (N) cytoplasm (Gracen *et al.*, 1971; Turner & Martinson, 1972).

METHODS AND MATERIALS

Initiation of cultures

Shoot-forming cultures of maize may be initiated from the scutella of immature embryos excised when they have reached a length of 1.0-2.0 mm (Green & Phillips, 1975). It is important to use a responsive genotype such as A188 to obtain a culture that will retain its shoot-forming capacity when maintained over a long period by serial subculture. Cultures carrying Tms cytoplasm (T-cultures) are produced by crossing a Texas male-sterile plant with pollen lacking dominant fertility restorer genes. Embryos are best isolated by cutting the kernels at the base and gently squeezing the milky endosperm. This allows the embryo to be picked out aseptically with tweezers and to be placed on the nutrient medium with the plumule-radicle axis in contact with the agar surface. A satisfactory response is obtained using the medium of Green & Phillips (1975) in a slightly modified form, consisting of Murashige & Skoog (1962) inorganic salts (revised medium) plus the following: glycine (7.7 mg/l); L-asparagine (2.0 g/l); nicotinic acid (1.3 mg/l); thiamine HCl (0.25 mg/l); pyridoxine HCl (0.25 mg/l); calcium pantothenate (0.25 mg/l); biotin (4.0 µg/l); 2,4-dichlorophenoxyacetic acid (2,4-D; 1.0 mg/l); sucrose (20.0 g/l); agar (8.0 g/l). The cultures have an approximate doubling time of 8 days. They are best grown in plastic Petri dishes (90 mm) sealed with Parafilm, incubated in the dark at 25°C, and maintained by transferring small pieces of tissue (50-60 mg fresh weight) every 6 weeks to fresh medium in which the concentration of 2,4-D is increased to 1.25 mg/l.

Preparation of toxin

Spores of *D. maydis* race T are obtained from dried infected leaves of maize ground gently in distilled water using a mortar and pestle. The spores are concentrated in the liquid fraction by slow speed centrifugation (250 *g*) and plated on potato dextrose agar (PDA). Single spore colonies are transferred to fresh PDA, from which T-toxin may be extracted after 15 days of incubation according to the procedure of Halloin *et al.* (1973). It has been observed that T-toxin is very heat-stable (Smedegard-Petersen & Nelson, 1969); therefore the aqueous toxin extract may be incorporated in the culture medium before autoclaving.

Determination of the sensitivity of the cultures to toxin

It is necessary to demonstrate that sensitivity to the toxin is expressed by the cultured tissue and to determine a suitable concentration of toxin for the selection of resistant lines. For this purpose, cultures are grown on media containing a wide range of concentrations of toxin. The results of such an experiment are given in Fig. 1. Small pieces of maize tissue (each approximately 50 mg) were placed on medium containing T-toxin and incubated for 9 weeks, after which time the fresh weights of the cultures were measured. In this example, the growth of cultures derived from embryos carrying normal (N) cytoplasm is not affected by the presence of 100 ml/l T-toxin in the culture medium, whereas the growth of cultures derived from embryos carrying Tms cytoplasm (T-cultures) is markedly inhibited by concentrations greater than 1.0 ml/l. At concentrations of 10 and 100 ml/l the T-cultures fail to make any further growth when subsequently transferred to toxin-free medium.

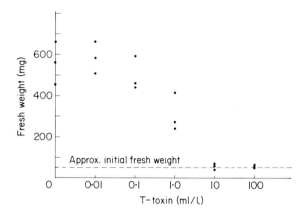

Figure 1. Fresh weight of maize cultures after 9 weeks on medium containing T-toxin.

Selection of resistant tissue

The technique of recurrent selection involves placing pieces of cultured tissue on medium containing a near-lethal concentration of toxin, and selecting out fast-growing viable sectors of tissue from the remainder of the cultured material killed by the toxin. This method has been successfully applied to maize for the isolation of lines resistant to T-toxin from originally susceptible T-cultures (Gengenbach & Green, 1975; Brettell *et al.*, 1979). Resistant tissue obtained in this way continues to proliferate on medium containing high levels of T-toxin. Further, the resistance is found to be stable in the absence of selecting toxin and is correlated with an increased resistance of isolated mitochondria to effects of the toxin which are associated with mitochondrial membrane function (Gengenbach & Green, 1975; Brettell *et al.*, 1979).

Regeneration of plants from cultures selected for toxin resistance

The next step in the procedure is to regenerate plants from the cultured material. In many instances the capacity of tissue cultures to regenerate plants is lost over a long period of continuous subculture (Reinert *et al.*, 1977). In maize, it is possible to maintain the shoot-forming ability of embryo-derived tissue cultures by taking care during subculture not to transfer the nodular mucilaginous tissue character-istic of a root-type callus which is generally incapable of plant formation (Mott & Cure, 1978). Regeneration is achieved by reducing the

level of 2,4-D in the medium to 0.25 mg/l and incubating the cultures
in light of 2000 lx (Green & Phillips, 1975). After 4 weeks the shoot
clusters that are formed may be transferred to the same medium without
2,4-D. Under these conditions roots are formed at the base of the
shoots, and after 10 days the resulting plantlets can be picked out of
the agar medium and placed in soil.

Testing the plants regenerated from tissue culture for toxin resistance

It is important that the plants obtained from cultures selected for
toxin resistance should be critically examined for expression and
inheritance of the resistance character. Only in this way can it be
determined whether the resistance observed *in vitro* is due to a
mutational event or to epigenetic adaptation by the cultured cells to
the presence of toxin. The plants regenerated from maize cultures
selected from originally toxin-sensitive Tms material are highly
resistant to T-toxin (applied to leaf tissue), but show a reversion to
male-fertility (Gengenbach *et al.*, 1977; Brettell *et al.*, 1980).
Further studies on the progeny of these plants have shown that male-
fertility and toxin resistance are inseparable and are inherited
together maternally.

The results are consistent with the toxin acting to select cells
enriched with T-toxin-resistant mitochondria similar to those found in
cells with normal (N) cytoplasm. Toxin-resistant, male-fertile maize
plants have, however, also been regenerated from T-cultures never
exposed to T-toxin. Of 60 plants produced from T-cultures maintained
for more than 1 year on agar medium, 31 were fertile and toxin-resistant
(Brettell *et al.*, 1980). This suggests that changes in the mitochondrial population contributing to a toxin-resistant tissue occur
spontaneously in culture, although the selection of a resistant tissue
is greatly facilitated by the addition of toxin to the culture medium.

The failure to separate toxin-sensitivity and Texas male-sterility
has meant that the plants resistant to T-toxin obtained by recurrent
selection are of no practical benefit to the plant breeder. However,
the results demonstrate the effectiveness with which a microbial toxin
can be used to isolate toxin-resistant cells in culture, and show that
plants regenerated from such cultures may express a heritable toxin
resistance. In other host-parasite systems where a toxin is a determinant of pathogenicity a more systematic approach may be necessary for the
induction of stable toxin-resistant mutant cells in culture. Nevertheless, it is clear that *in vitro* selection techniques would provide a
feasible means of obtaining novel disease-resistant plants.

Acknowledgements

D.S. Ingram acknowledges the receipt of an Agricultural Research Council
support grant. Work on *D. maydis* in the UK was carried out under MAFF
licence no. 11414/169 issued under the Destructive Pests and Diseases of
Plants Order, 1965.

REFERENCES

Brettell R.I.S., Goddard B.V.D. & Ingram D.S. (1979) Selection of Tms-
    cytoplasm maize tissue cultures resistant to *Drechslera maydis*
    T-toxin. *Maydica* 24, 203-13.
Brettell R.I.S., Thomas E. & Ingram D.S. (1980) Reversion of Texas
    male-sterile cytoplasm maize in culture to give fertile, T-toxin
    resistant plants. *Theoretical and Applied Genetics* (in press).
Gengenbach B.G. & Green C.E. (1975) Selection of T-cytoplasm maize
    callus cultures resistant to *Helminthosporium maydis* race T patho-
    toxin. *Crop Science* 15, 645-9.
Gengenbach B.G., Green C.E. & Donovan C.M. (1977) Inheritance of
    selected pathotoxin resistance in maize plants regenerated from cell
    cultures. *Proceedings of the National Academy of Sciences, USA* 74,
    5113-7.
Gracen V.E., Forster M.J., Sayre K.D. & Grogan C.O. (1971) Rapid method
    for selecting resistant plants for control of southern corn leaf
    blight. *Plant Disease Reporter* 55, 469-70.
Green C.E. & Phillips R.L. (1975) Plant regeneration from tissue
    cultures of maize. *Crop Science* 15, 417-21.
Halloin J.M., Comstock J.C., Martinson C.A. & Tipton C.L. (1973)
    Leakage from corn tissues induced by *Helminthosporium maydis* race T
    toxin. *Phytopathology* 63, 640-2.
Lim S.M. & Hooker A.L. (1972) Disease determinant of *Helminthosporium*
    maydis race T. *Phytopathology* 62, 968-71.
Maliga P., Sz. Breznovits A. & Márton L. (1973) Streptomycin-resistant
    plants from callus culture of haploid tobacco. *Nature New Biology*
    244, 29-30.
Matern U., Strobel G. & Shepard J. (1978) Reaction to phytotoxins in a
    potato population derived from mesophyll protoplasts. *Proceedings of*
    the National Academy of Sciences, USA 75, 4935-9.
Mott R.L. & Cure W.W. (1978) Anatomy of maize tissue cultures.
    Physiologia Plantarum 42, 139-45.
Murashige T. & Skoog F. (1962) A revised medium for rapid growth and
    bioassays with tobacco tissue cultures. *Physiologia Plantarum* 15,
    473-97.
Patil S.S. (1974) Toxins produced by phytopathogenic bacteria. *Annual*
    Review of Phytopathology 12, 259-79.
Pringle R.B. & Scheffer R.P. (1964) Host specific plant toxins. *Annual*
    Review of Phytopathology 2, 133-56.
Reinert J., Bajaj Y.P.S. & Zbell B. (1977) Aspects of organisation -
    organogenesis, embryogenesis, cytodifferentiation. *Plant Tissue and*
    Cell Culture (Ed. by H.E. Street), pp. 389-427. Blackwell Scientific
    Publications, Oxford.
Smedegard-Petersen V. & Nelson R.R. (1969) The production of a host-
    specific pathotoxin by *Cochliobolus heterostrophus*. *Canadian Journal*
    of Botany 47, 951-7.
Turner M.T. & Martinson C.A. (1972) Susceptibility of corn lines to
    Helminthosporium maydis toxin. *Plant Disease Reporter* 56, 29-32.

Methods for selection of drug-resistant plant cell cultures

A. STRAUSS, C. GEBHARDT & P.J. KING
Friedrich Miescher-Institut, PO Box 273,
CH-4002 Basel, Switzerland

INTRODUCTION

For many years the idea has existed that large populations of single
somatic cells from higher plants might be handled like micro-organisms
and used to search in a precise way for biochemical mutants. Especially
interesting is the notion that mutant plants can be regenerated from
mutant cell populations and that this technique might offer a new, more
effective way of screening for such agronomically important phenotypes
as disease resistance, resistance to herbicides and crop protectants,
resistance to suboptimal physical growth conditions and improved
nutritional quality. However, it is only recently that a few plant cell
culture and protoplast systems have been perfected to the point where
they can be used for somatic cell genetics, and the failure to regenerate plants from putative mutant cells has been particularly frustrating.

The application of *in vitro* culture techniques to the selection of
disease-resistant plants (see also Brettell & Ingram, 1979) is restricted for the moment to those diseases caused by the toxins of pathogenic
micro-organisms. The successful use of such methods has been reported
by Carlson (1973), who obtained simultaneous resistance of *Nicotiana
tabacum* to *Pseudomonas tabaci* and methionine sulfoximine, and by
Gengenbach *et al.* (1977) and Brettell *et al.* (this volume), who obtained
simultaneous resistance of *Zea mays* to *Helminthosporium maydis* and race
T toxin. As techniques for the direct infection of plant tissue
cultures with pathogenic micro-organisms have already been reported
(Helgeson *et al.*, 1976) it should also be possible to select *in vitro*
for disease-resistant cultures and to regenerate plants resistant to
diseases where no pathotoxin is involved. On the other hand, plants
with a disease resistance which depends on a structural attribute (e.g.
cuticle thickness) are not likely to be recovered from selection
experiments with cell cultures.

The advantage of *in vitro* culture systems over conventional breeding
procedures for obtaining resistant mutant plants lies in the fact that
a large number of individuals can be screened for a resistance character
in a rapid, space-saving and controlled way under constant conditions.

Many kinds of plant material, such as plants, seeds, pollen, embryos, explants, tissue cultures, calluses, cell suspensions and protoplasts, growing in sterile laboratory conditions, can be used for selection of resistant variants. Only a few of these, however, can be handled in such a way that the potential advantages of selection for resistant individuals among a huge population of totipotent cells can be fully exploited. For example, cell cultures of the "cell line" type (King *et al.*, 1978), indeed represent homogeneous cell populations but appear to have lost their morphogenic capacity and are thus only valuable as model systems. Unfortunately, many other cultures growing on solid medium are merely proliferating explants consisting of a mixture of different tissues containing only a limited number of totipotent cells (King *et al.*, 1978). Some of the tissue culture problems related to crop improvement have been discussed by Thomas *et al.* (1979).

   The ideal attributes of a system for selecting resistant mutants from plant cultures *in vitro* are as follows.
   1. Single cells (preferably haploid).
   2. Simple, short, inexpensive and reproducible isolation procedures.
   3. Routine regeneration of resistant plants.
   4. Indefinite and simple storage.
   5. A complete, stable and heritable resistance character without
      pleiotropic effects.
   6. A valuable character.
   7. A known genetic and biochemical basis for the resistance character.
Although this situation has not as yet been totally achieved, the systems which approximate most closely to it are those using mesophyll protoplasts from solanaceous species (e.g. Bourgin, 1978).

   In this article we outline the various theoretical and practical problems involved in the search for drug-resistant plants and cultures using *in vitro* techniques. Drug resistance stands as a model example for resistance, and many similar problems occur in selection for other kinds of resistance. Every system in which selection of drug-resistant variants is attempted will pose new and specific problems. As methods for selection for resistance at the tissue culture level have already been discussed by Brettell *et al.* (this volume) we shall mainly consider aspects of the use of cell cultures.

METHODS AND MATERIALS

Cell cultures

The general methods for initiation and care of plant cell cultures, as well as their regeneration into plants, are given in several textbooks (e.g. Street, 1977).

   It is assumed that the starting material (callus, cell suspension culture, protoplasts) is free of infection by micro-organisms and represents a homogeneous population with respect to sensitivity to the drug in question. To avoid a high proportion of chimeras after

selection, single cells with high division potential are preferred.
Multicellular units can, however, also be employed because in any case
purification of the resistant cell clones after isolation is inevitable.
The cultures should be morphogenic.

Truly haploid cell material offers the broadest possible spectrum of
resistant mutants, as both recessive and dominant traits can be
expressed at this ploidy level. Dominant and semi-dominant characters
can also be successfully selected from cells of a diploid or even a
higher ploidy level. Dominant resistance is expected, for instance,
when an antimetabolite-resistance character is due to overproduction of
the corresponding metabolite. It should also be possible to obtain
recessive mutants from diploid cell material by induction of mitotic
recombination subsequent to mutagenization, although this has never been
reported for plant cell cultures. In this procedure a recessive
mutation in the heterozygous state would eventually become homozygous
and could thus be expressed. Induction of mitotic recombination by
various agents has been reported for different plant species (e.g. Evans
& Paddock, 1977).

Mutagenization

Mutagenization by chemicals or irradiation has not yet been quantified
for any plant cell culture system as has been done for whole plants
(Ehrenberg, 1971; Coe & Neuffer, 1977) and mammalian cell cultures
(O'Neill & Hsie, 1977). However, some reports suggest that mutant
frequencies are increased in cultured plant cells (Malepszy *et al.*,
1977; Widholm, 1977; Colijn *et al.*, 1979). It is reasonable to assume
that the DNA carrying the genetic information of plant cells in culture
is modified by mutagen treatment in ways similar to that documented
for whole plants and other organisms. Nevertheless, differences in
susceptibility to mutagens might exist between different systems because
of differences in the cellular environment influencing mutagenization
and in DNA repair mechanisms.

The most commonly used mutagens for plants are N-methyl-N'-nitro-
N-nitrosoguanidine (MNNG), ethyl methanesulfonate (EMS), sodium azide,
X-rays and UV-rays. The right choice of the mutagen depends on the kind
of mutation desired (point mutations, larger chromosomal alterations
etc.) and on the state of the cells to be mutagenized (resting,
dividing). For detailed information about mutagens and their action
see, for example, the reviews of Freese (1971) and Gottschalk (1975).
Until optimum conditions for mutagenization of plant cell cultures are
known, inactivation has to be taken as a measure of mutagen efficiency.
Usually mutagen doses giving 10-90% survival are applied. From non-
plant systems it is known that in general the mutant fraction of the
survivors increases with the mutagen dose (Fincham & Day, 1971; O'Neill
& Hsie, 1977), although Colijn *et al.* (1979) found a high number of
resistant plant cell variants using non-lethal rather than lethal doses
of MNNG. A potent mutagen should increase the spontaneous mutation
rate, usually in the order of 10^{-8} to 10^{-6} mutations per cell per
generation for a given drug-resistance character, by a factor of more

than 100.

Many chemical mutagens are unstable (Ehrenberg & Wachtmeister, 1977) and should therefore be freshly prepared and applied only for a short period under constant conditions of buffer, pH, time, temperature, light intensity, cell density and state of cells. All mutagens represent a serious risk to the health of the investigators and special safety precautions should therefore be taken in handling, applying and disposing of them.

Drug-resistant variants from plant cell cultures have sometimes been obtained without any mutagenization (e.g. Widholm, 1977). Many drugs are themselves mutagenic. Also, since positive selection for drug resistance allows the screening of millions of cellular units at once, it is obviously possible to recover spontaneously occurring resistant clones.

Selection

Before performing a selection experiment it is necessary to know the drug sensitivity of the wild type cell material under standard selection conditions. This is most appropriately done by establishing a dose-response curve. The inactivating effect of a drug on cells can be assessed by determination either of viability, using such characters as plating efficiency, dye uptake, cytoplasmic streaming and incorporation of labelled metabolites into macromolecules, or of growth, by determining fresh weight or dry weight (Street, 1977). Unfortunately, all these criteria have their drawbacks and it is therefore inadvisable to rely solely on any one method. Where large cell masses are involved and penetration of a drug into deeper cell layers plays a role, as with big callus pieces, it is necessary to carry the culture through a second passage in the presence of the drug. The cell density in dose-response determinations must be the same as in the selection experiments, as drug sensitivity is strongly dependent upon the mass of drug-detoxifying substances present in the cells.

As with microbial and mammalian cell culture systems (Moyed, 1964; O'Neill & Hsie, 1977), it can be assumed that an expression time between mutagenization and drug application is required for successful selection. It is generally thought that this expression time should be several cell generations, to dilute out enzymes which render the mutant cells sensitive despite their altered genetic information.

Selection is best carried out at a drug concentration which will either kill or totally inhibit all wild type cells. Doses which are too low should be avoided because cells escaping for purely physiological reasons could outnumber true mutants. On the other hand, drug concentrations which are too high may mean that potentially resistant variants are also unable to grow and are thus lost. Cells should be incubated in a drug-containing medium for a sufficient length of time to ensure expression and growth of the resistant clones.

It has often been demonstrated (Street, 1977) that growth of plant cells *in vitro* at low cell density is dependent on the presence of growth promoting (conditioning) factors in the culture medium. This low cell density effect can usually be overcome experimentally, for example by the application of feeder cells (Raveh et al., 1973; Weber & Lark, 1979) or by supplementation with organic compounds (Kao & Michayluk, 1975). Further, it is known, at least for protoplast systems (Street, 1977), that growth is inhibited at very high cell densities. Dividing, arrested, dying and dead cells could well differ in release and adsorption of compounds affecting growth. In other words, the optimal cell density of drug-treated and non-treated cells could be totally different. Before making selection experiments for drug resistance, it is thus important to know whether a few variant cell clones would have any chance of surviving among many inactivated cells. Reconstitution experiments, for instance, with mixtures of different proportions of living and drug-inactivated cell material, could help to determine whether the cell density is critical under the conditions of selection (Weber & Lark, 1979; Strauss & King, unpublished data).

The following procedures have been used with success for the selection of drug-resistant variants from plant cell or tissue cultures.
1. Recurrent selection using callus or tissue pieces on drug-containing agar media (e.g. Maliga et al., 1973; Gengenbach et al., 1977).
2. Selection by plating freshly dispersed callus or tissue pieces in drug-containing agar media (e.g. Müller & Grafe, 1978).
3. Selection by plating established cell suspension cultures in drug-containing agar media (Chaleff & Parsons, 1978).
4. Selection by incubating established cell suspension cultures in shaken liquid media in the presence of the drug (e.g. Widholm, 1976).
5. Selection by incubating protoplast-derived colonies in drug-containing media (e.g. Bourgin, 1978).

To our knowledge, the successful application of non-selective techniques for the isolation of resistant variants from plant *in vitro* cultures (sib-selection, total isolation) has not been reported.

Isolation and purification

Contamination of drug-resistant clones with non-resistant cell material may occur because:
1. the resistant clones were picked off together with arrested wild type cells from the immediate surroundings, which subsequently recovered on drug-free medium;
2. the cellular units subjected to mutagenization were multicellular or multinuclear or in the G2 phase of the cell cycle;
3. mutations were fixed in only one of the two complementary DNA strands and were transferred to only one of the two daughter cells, as often happens in fungi (Fincham & Day, 1971);
4. the resistant trait was highly unstable (rare).

It is therefore necessary to eliminate the contaminating wild type cell material during the isolation and purification procedure. This can be

done by picking off growing callus sectors, tissue sectors or colonies
from drug-containing medium as cleanly as possible, and subculturing
them for some passages on selective medium. Purification is most
reliable when done by cell cloning via protoplasts or by plant regeneration and subsequent crossings. If neither technique is applicable, the
resistant cell material should be subcultured so that the chances of
ending with a pure culture of resistant cells are increased; this can be
achieved by repeatedly subculturing small pieces of callus on selective
medium, or by isolating new single colonies from the replated resistant
culture.

Confirmation

The resistant cell clones have to be cultured for further increase in
mass of cell material to occur, and should be converted into the same
type of culture as the original wild type strain. As soon as the cell
material of the resistant variants is ready, its growth or viability on
drug-containing and drug-free media should be quantitatively compared
with that of the wild type strain, using dose-response curves. In this
way the increase in resistance of the resistant variants compared with
the wild type can be assessed, and undesirable strains such as highly
unstable resistant variants, physiological escapes and variants with an
altered growth pattern (e.g. with the same relative increase in growth
on media with or without drugs) can be detected. If growth is used to
evaluate drug sensitivity, the growth pattern (lag phase, growth rate
and growth yield) of the resistant variants must first be compared with
that of the wild type strain. Differences in growth pattern can lead
to false interpretation of dose-response curves, and thus the timing of
growth measurements is critical (Nielsen *et al.*, 1979). Note that a
change in sensitivity cannot be excluded despite a stable genetic basis
for resistance when confirmation of the resistant phenotype takes place
in a type of culture different from that in which the selection was
performed.

Some compounds are known to induce a transient phenotypic resistance
in fungi (Moyed, 1964) and in plant cell cultures (Zryd, 1979). Such
events can be recognized by retesting the degree of resistance of the
variants after prolonged incubation on drug-free medium. The stability
of the resistant phenotype under non-selective conditions has often been
taken to indicate a change of genetic origin, although some doubts still
remain (Maliga, 1976).

Storage

Any resistant variant becomes useless if it is lost through microbial
infection, reversion to sensitivity or some mistake in subculturing.
It is therefore important to have a reliable method for storing the
cultures before the selection of resistant variants is attempted. If
long term storage of seeds is possible, the resistant character is most
conveniently stored in the form of seeds from regenerated plants. In
the last few years deep-freezing methods for use with plant cell and
tissue cultures have been developed (Withers, 1980), but species- and

culture-specific storage protocols have to be devised before such methods can be routinely adopted. If neither of these two storage methods is feasible, resistant variants should be maintained by subculturing, preferably in independent parallel cultures on selective medium.

Classification and characterization

In most cases a series of different resistant variants is isolated at once. Depending on the reasons for the search for resistant variants, classification and further characterization of the variants is interesting or even necessary.

In a first round of classification the kind of genetic change causing the drug-resistant phenotype is determined for different variants. Where plant regeneration is possible, the methods of classical genetics are used to decide if one is dealing with nuclear mutations, cytoplasmic mutations or epigenetic variations. Chromosomal aberrations can be detected by means of cytogenetic techniques, even in the case of non-morphogenic cultures. More information about the nature of the genetic change can be obtained through protoplast fusion techniques, estimation of forward and reverse mutation rates and direct biochemical analyses of structural and functional changes in proteins or nucleic acids.

In a second round of genetic classification the localization of the genetic defect on the genome and the dominance-recessiveness relationships are established. This is done by recombination and complementation analyses using sexual crosses with regenerated plants. While the parasexual cycle remains impracticable with plant cell cultures, *in situ* DNA hybridization techniques and complementation studies using protoplast fusion remain the only theoretically possible means of such a classification of non-morphogenic resistant variants.

Initial information on the biochemical basis for resistance can be obtained by a phenotypic characterization of the resistant variants, including tests for cross-resistance to other drugs or a search for pleiotropic effects on cell morphology etc. Direct biochemical analysis of the variants may reveal alterations in the structure or function of enzymes, and changes in uptake, localization, synthesis, turnover and pool size of metabolites and antimetabolites.

CONCLUSIONS

We have tried to summarize the principle steps in the isolation of resistant mutants from plant somatic cell populations. Many of the points we make are as yet only theoretical, while others will seem obvious to those with experience either of plant tissue culture or genetics. It is, however, already clear that many tissue cultures are far from ideal for somatic cell genetics. Therefore, when devising a strategy for mutant isolation, the first essential is to become familiar with the form and capabilities of the cell system to be used, so that many artefacts can be avoided.

Mendelian genetics has a long history; by comparison, plant tissue culture is still in a primitive state. It is, however, very encouraging that some recently developed culture systems meet most of the requirements for successful mutant isolation (e.g. Bourgin, 1978; Wernicke et al., 1979).

Acknowledgement

We are grateful to colleagues at the FMI for helpful discussions.

REFERENCES

Bourgin J.-P. (1978) Valine-resistant plants from *in vitro* selected tobacco cells. *Molecular and General Genetics* 161, 225-30.

Brettell R.I.S. & Ingram D.S. (1979) Tissue culture in the production of novel disease-resistant crop plants. *Biological Reviews* 54, 329-45.

Carlson P.S. (1973) Methionine sulfoximine-resistant mutants of tobacco. *Science* 180, 1366-8.

Chaleff R.S. & Parsons M.F. (1978) Direct selection *in vitro* for herbicide-resistant mutants of *Nicotiana tabacum*. *Proceedings of the National Academy of Sciences, USA* 75, 5104-7.

Coe E.H., Jr. & Neuffer M.G. (1977) The genetics of corn. *Corn and Corn Improvement* (Ed. by G.F. Sprague), pp. 111-223. American Society of Agronomy, Inc., Madison, Wisconsin.

Colijn C.M., Kool A.J. & Nijkamp H.J.J. (1979) An effective chemical mutagenesis procedure for *Petunia hybrida* cell suspension cultures. *Theoretical and Applied Genetics* 55, 101-6.

Ehrenberg L. (1971) Higher plants. *Chemical Mutagens. Principles and Methods for their Detection* (Ed. by A. Hollaender) Vol. 2, pp. 365-86. Plenum Press, New York & London.

Ehrenberg L. & Wachtmeister C.A. (1977) Handling of mutagenic chemicals: Experimental safety. *Handbook of Mutagenicity Test Procedures* (Ed. by B.J. Kilbey), pp. 411-8. Elsevier, Amsterdam.

Evans D.A. & Paddock E.F. (1977) X-ray induced increase of mitotic crossovers in *Nicotiana tabacum*. *Environmental and Experimental Botany* 17, 99-106.

Fincham J.R.S. & Day P.R. (1971) The induction, isolation and characterization of mutants. *Fungal Genetics* (Ed. by J.H. Burnett), 3rd edition, pp. 46-85. Blackwell Scientific Publications, Oxford.

Freese E. (1971) Molecular mechanisms of mutations. *Chemical Mutagens. Principles and Methods for their Detection* (Ed. by A. Hollaender), Vol. 1, pp. 1-56. Plenum Press, New York & London.

Gengenbach B.G., Green C.E. & Donovan C.M. (1977) Inheritance of selected pathotoxin resistance in maize plants regenerated from cell cultures. *Proceedings of the National Academy of Sciences, USA* 74, 5113-7.

Gottschalk W. (1975) Mutation. *Fortschritte der Botanik* 37, 219-46.

Helgeson J.P., Haberlach G.T. & Upper C.D. (1976) A dominant gene conferring disease resistance to tobacco plants is expressed in tissue cultures. *Phytopathology* 66, 91-6.

Kao K.N. & Michayluk M.R. (1975) Nutritional requirements for growth of *Vicia hajastana* cells and protoplasts at very low population density in liquid media. *Planta* 126, 105-10.

King P.J., Potrykus I. & Thomas E. (1978) *In vitro* genetics of cereals: problems and perspectives. *Physiologie Végétale* 16, 381-99.

Malepszy S., Grunewaldt J. & Maluszynski M. (1977) Ueber die Selektion von Mutanten in Zellkulturen aus haploider *Nicotiana sylvestris* Spegazz. et Comes. *Zeitschrift für Pflanzenzüchtung* 79, 160-6.

Maliga P. (1976) Isolation of mutants from cultured plant cells. *Cell Genetics in Higher Plants* (Ed. by D. Dudits, G.L. Farkas & P. Maliga), pp. 59-76. Akadémiai Kiadó, Budapest.

Maliga P., Sz.-Breznovits A. & Márton L. (1973) Streptomycin-resistant plants from callus culture of haploid tobacco. *Nature New Biology* 244, 29-30.

Moyed H.S. (1964) Biochemical mechanisms of drug resistance. *Annual Review of Microbiology* 18, 347-64.

Müller A.J. & Grafe R. (1978) Isolation and characterization of cell lines of *Nicotiana tabacum* lacking nitrate reductase. *Molecular and General Genetics* 161, 67-76.

Nielsen E., Rollo F., Barisi B., Cella R. & Sala F. (1979) Genetic markers in cultured plant cells: differential sensitivities to amethopterin, azetidine-2-carboxylic acid and hydroxyurea. *Plant Science Letters* 15, 113-25.

O'Neill J.P. & Hsie A.W. (1977) Chemical mutagenesis of mammalian cells can be quantified. *Nature* 269, 815-7.

Raveh D., Huberman E. & Galun E. (1973) *In vitro* culture of tobacco protoplasts: use of feeder techniques to support division of cells plated at low densities. *In Vitro* 9, 216-22.

Street H.E. (1977) *Plant Tissue and Cell Culture*, 2nd edition. Blackwell Scientific Publications, Oxford.

Thomas E., King P.J. & Potrykus I. (1979) Improvement of crop plants via single cells *in vitro* - an assessment. *Zeitschrift für Pflanzenzüchtung* 82, 1-30.

Weber G. & Lark K.G. (1979) An efficient plating system for rapid isolation of mutants from plant cell suspensions. *Theoretical and Applied Genetics* 55, 81-6.

Wernicke W., Lörz H. & Thomas E. (1979) Plant regeneration from leaf protoplasts of haploid *Hyoscyamus muticus* L. produced via anther culture. *Plant Science Letters* 15, 239-49.

Widholm J.M. (1976) Selection and characterization of cultured carrot and tobacco cells resistant to lysine, methionine, and proline analogs. *Canadian Journal of Botany* 54, 1523-9.

Widholm J.M. (1977) Isolation of biochemical mutants of cultured plant cells. *Molecular Genetic Modification of Eukaryotes* (Ed. by I. Rubinstein, R.L. Phillips, C.E. Green & R.J. Desnick), pp. 57-65. Academic Press, New York.

Withers L.A. (1980) Low temperature storage of plant tissue cultures. *Advances in Biochemical Engineering* 18, 101-50.

Zryd J.P. (1979) Colchicine-induced resistance to antibiotic and amino-acid analogue in plant cell cultures. *Experientia* 35, 1168-9.

A strategy for the production of disease-resistant mutants

B.W.W. GROUT* & M.A. WEATHERHEAD†
*Department of Biology, North East London Polytechnic,
Romford Road, London E15 4LZ
†Department of Plant Sciences, University of Leeds, LS2 9JT

INTRODUCTION

The most efficient way of controlling plant disease is through the
identification and utilization of resistant varieties. Natural sources
of resistance lie in the genetic diversity contained in existing land
races, varieties and wild species, and as far as possible these must be
screened for the presence of potentially useful characters. However,
when natural sources of resistance cannot be identified the possibility
of using artificially induced mutants must be explored. The acceler-
ating depletion of genetic diversity in many species will undoubtedly
increase the future importance of this approach. There are, however,
difficulties to be faced in successful mutant production. Firstly the
frequency of mutation may be low and thus large quantities of material
must be treated if useful mutants are to be recovered, and secondly
induced mutations in diploid (or polyploid) tissues are commonly
recessive and therefore not expressed in the phenotype. Further, even
when a potentially useful mutant is isolated it has to be propagated
for several generations before a population large enough for continued
trials can be produced.

In an attempt to simplify the induction and identification of
mutations we have compiled a strategy aimed at the production of
disease-resistant plants by drawing on proven techniques from several
disciplines. We include mutagenic treatments from the field of plant
breeding, disease screening methods from the field of plant pathology
and a variety of tissue culture techniques. The strategy consists of
the following procedures: selection of haploid material (i.e. pollen
grains); exposure to mutagenic agents; induction of haploid plants in
anther/pollen culture; chromosome doubling to recover homozygous
diploid plants; screening of diploids for disease-resistant mutants;
clonal micropropagation of selected mutants in tissue culture; and
confirmatory trials and further propagation.

In order to develop a model system we are currently using
Saintpaulia ionantha as experimental material, and screening for
induced mutations conferring improved resistance to fungal pathogens

Ingram D.S. & Helgeson J.P. (1980) *Tissue Culture Methods for Plant Pathologists*

such as *Pythium* spp., and improved cold hardiness. In the following
sections we discuss general considerations and describe specific
methods applicable to *S. ionantha*.

METHODS

Selection of haploid material

Several factors may influence the selection of suitable material for
successful anther or pollen culture, including the physiological status
of the parent plants. Pollen response can be affected by the age of
the plants (Sunderland, 1971), seasonal variation (Dunwell & Sunderland,
1973) and photoperiod (Dunwell & Perry, 1973), and therefore optimum
conditions have to be determined empirically. The developmental stage
of the pollen at the initiation of the culture is also of crucial
importance, most species responding only within a limited range of
development. In some species certain pretreatments (e.g. chilling)
have been shown greatly to increase yields of haploid plants (Sunderland
& Roberts, 1977) and where applicable these pretreatments should be
utilized to obtain optimum success in culture.

     Plants of *S. ionantha* should be maintained at 25-27°C in a 16 h
photoperiod, supplied artificially when necessary. Ideally anthers
should be excised only from young plants, but it is our experience that
the pollen remains responsive even when plants have been flowering for
several months, provided that plant vigour has been maintained by
regular applications of liquid fertilizer. For initiation of cultures
anthers should contain pollen at the late uninucleate stage of development, and identification of such anthers can be facilitated by prior
correlation of pollen development with morphological features such as
bud length or anther colour. As yet pretreatments of the anthers have
not resulted in an increased response in *S. ionantha*.

Exposure to mutagenic agents

Chemical mutagens   Detailed information concerning a wide range of
these compounds can be found in the Manual on Mutation Breeding (Anon.,
1977), which includes data on their modes of action and suggested
methods of application. The selection of a compound at a concentration
which is not toxic during the time course of the treatment, and which
does not significantly impair subsequent growth, is vital. To treat
pollen grains, isolated anthers can be soaked in a solution of the
mutagen or alternatively cultured for a limited period on growth medium
to which the mutagen has been added. Successful mutation then depends
on penetration and uptake of the mutagen by the haploid cells within
the anther. Similar treatments of somatic plant tissue cultures have
been found to be satisfactory (Heinz *et al.*, 1977; Maliga, 1978).

Physical mutagens   These are preferred for their ease of application,
good penetration, reproducibility and higher mutation frequency as
compared with chemical mutagens. Hard X-rays and gamma rays are the

most suitable types of ionizing radiation, with typical doses ranging
from 1-10 kilorad given at rates of 0.1-1.0 krad/min (Anon., 1974;
Howland & Hart, 1977; Broertjes & van Harten, 1978). Radiosensitivity
may vary greatly between different plant genotypes and also between
different plant organs, so the optimal dosage must be determined
empirically. The required dose is that which produces the maximum
number of mutations without impairing survival of the tissues.

At the time of writing our investigations have been restricted to
the use of physical mutagens and a low number of variants have been
induced by irradiation with gamma rays to give a total of 5580 rad
presented at 17.28 krad/h.

Induction of haploid plants

During anther and pollen culture a proportion of the haploid cells are
diverted from a gametophytic pathway of development onto a sporophytic
one to produce haploid plants either direct from a single cell
(embryogenesis) or by organogenesis from a pollen derived callus. To
ensure plantlet production from the maximum number of mutated cells
optimum culture conditions must be employed. A wide range of growth
substances and medium additives (e.g. activated charcoal; Anagnostakis,
1974) may have to be screened for the best results, the most suitable
combinations often being species-specific.

Plantlets that have arisen from pollen by embryogenesis can immediately be taken further in the strategy, and should contain a number of
potentially valuable mutants. Where shoot production is initiated by
organogenesis of a pollen derived callus, and shoots arise from a
single cell, such mutant plantlets as occur will be solid, but where
shoot induction involves a number of cells chimeral mutants may be
produced which are likely to be unsuitable for disease resistance
studies. The avoidance, where possible, of a callus stage is therefore
desirable.

Our studies of anther culture of *S. ionantha* to date have shown that
pollen derived callus may be induced with high frequency in anthers
cultured on Murashige & Skoog (MS; 1962) medium supplemented with
1 mg/l 1-naphthaleneacetic acid (NAA) and 0.5 mg/l 6-benzylaminopurine
(BAP). The callus, which is evident after some 4 weeks in culture, is
of high regenerative capacity, and when it is excised from the somatic
tissues and recultured on fresh medium of the same composition shoots
are formed within 2 to 4 weeks. Root induction can be achieved by
transference of the shoots to unsupplemented MS medium, after which
the plantlets may be taken into soil.

Although the callus route is not the most satisfactory method for
achieving haploidy we have demonstrated that some 200 plantlets may
be recovered from a single anther and that the majority of these are
haploid. Further studies are in progress to determine the origin of
these plantlets cytologically and to achieve direct embryogenesis of
the pollen grains either in anther or free pollen culture.

Chromosome doubling

Haploid plants of many species tend to lack vigour and may not therefore be suitable for disease resistance screening. To circumvent this
problem chromosome doubling agents can be applied to the plantlets,
which will give rise to homozygous diploid individuals. A convenient
method of achieving chromosome doubling is the *in vitro* application of
colchicine, a spindle arresting agent. Shoot tips are excised from
plantlets arising from anther culture and transferred to MS medium
containing 0.5% colchicine for 48 h, then transferred to unsupplemented
medium where whole plants are reformed. Ploidy levels of the haploids
and putative diploids are readily confirmed by root tip cytology.

We have found this technique to be successful for the doubling of
chromosomes in *Solanum* spp. (Weatherhead & Henshaw, 1979) and early
indications imply that it is also applicable to *S. ionantha*.

Screening for resistance

The method of exposure of the plantlet population to the pathogen will
depend upon the nature of the pathogen. If a biotroph is being tested
the plantlets can be maintained in tissue culture, whereas testing of
a necrotroph necessitates transfer of the population to sterile compost.
The method of application of the pathogen can also take several forms:
for instance drops of a spore suspension applied directly to leaves,
dry spores dusted over a plant, a spore suspension sprayed over a plant
(Crute & Dickenson, 1976), or incorporation of a known fungal phytotoxin into the growth medium. After a limited incubation period
(e.g. 2 weeks) those individuals which do not sustain a heavy infection
of the pathogen or which are obviously slow to develop symptoms are
selected for further testing.

When an *in vitro* screening method is being used due consideration
must be paid to the fact that the non-expression of resistance genes
in culture has been recorded for several species (Brettell & Ingram,
1979). If such non-expression is suspected the plant population must
be made physiologically independent by being transferred to sterile
soil and screened using the method described above for necrotrophic
pathogens, and used by Crute & Davis (1977) for screening seedlings
of *Lactuca sativa* for resistance to *Bremia lactucae*. Transplanting,
however, constitutes a major risk as anything less than 100% success
may mean the loss of valuable mutants (Grout & Crisp, 1977).

Among the several fungal pathogens of *S. ionantha* are *Pythium* spp.,
and we intend to screen plant populations for improved resistance to
these pathogens in the following stages: transferring mutant plantlets
from culture into sterile soil and allowing them to become well
established; transferring the plants into soil known to be infected
with oospores of *Pythium* spp. (ideally this soil should come from
commercial glasshouses where the presence of *Pythium* spp. has been
confirmed); and selecting resistant individuals to be taken back into
culture for cloning.

Clonal micropropagation of selected mutants

Individuals selected by screening can be rapidly multiplied for further testing by being taken back into tissue culture. The most suitable explants for propagation tend to be shoot-tip meristems, as a result of their genetic stability in culture, and from such explants a clone of several thousand genetically identical plants can be propagated within 1 year (Westcott *et al.*, 1979). For some species other plant organs may be more suitable, but in all cases regeneration via a callus phase should if possible be avoided as there is a risk of genetic alteration which may damage or destroy mutant properties.

S. ionantha is particularly amenable to micropropagation techniques and most organs and tissues exhibit high regenerative capacities. In terms of plantlet recovery we consider leaf discs to be the most suitable explant, since when these are cultured on MS medium containing 1 mg/l each of NAA and BAP numerous plantlets arise direct from the epidermal cells. A limited amount of callus may be produced at the cut surfaces of the discs, but any plantlets arising from this can be identified by inspection and discarded. Using this technique up to 150 plantlets can be produced from a single leaf disc in a 6-week culture period.

Confirmatory trials

The mutant clones can now be screened to determine the nature and extent of their enhanced disease resistance, and their other varietal characteristics can be checked. The selected clones, being homozygous for the genes regulating their resistance characters, can readily be incorporated into breeding programmes or subjected to further variety trials.

CONCLUSION

Using such a strategy it should be possible, assuming that the tissue culture techniques are readily available and that the required mutations occur, to produce large numbers of a homozygous disease-resistant mutant within 12-18 months of beginning the anther culture procedures.

Acknowledgement

The authors wish to thank Dr K.C. Short and Mr J. Kokhar (NELP) for access to, and discussion of, unpublished data.

REFERENCES

Anagnostakis S.L. (1974) Haploid plants from anthers of tobacco -
    enhancement with charcoal. *Planta* 115, 281-3.

Anon. (1974) *Induced Mutations for Disease Resistance in Crop Plants*.
    International Atomic Energy Authority, Vienna.
Anon. (1977) *Manual on Mutation Breeding*. Technical Report Series 119.
    International Atomic Energy Authority, Vienna.
Brettell R.I.S. & Ingram D.S. (1979) Tissue culture in the production
    of novel disease-resistant crop plants. *Biological Reviews* 54,
    329-45.
Broertjes C. & van Harten A.M. (1978) *Application of Mutation Breeding
    Methods in the Improvement of Vegetatively Propagated Crops*.
    Elsevier, Amsterdam.
Crute I.R. & Davis A.A. (1977) Specificity of *Bremia lactucae* from
    Lactuca sativa. *Transactions of the British Mycological Society*
    69, 405-10.
Crute I.R. & Dickenson C.H. (1976) The behaviour of *Bremia lactucae*
    on cultivars of *Lactuca sativa* and other composites. *Annals of
    Applied Biology* 82, 433-50.
Dunwell J.M. & Perry M.E. (1973) The influence of *in vivo* growth
    conditions of *N. tabacum* plants on the *in vitro* embryogenic
    potential of their anthers. *Annual Report of the John Innes Research
    Institute, No.* 64.
Dunwell J.M. & Sunderland N. (1973) Anther culture of *Solanum
    tuberosum* L. *Euphytica* 22, 317-23.
Grout B.W.W. & Crisp P.C. (1977) Practical aspects of the propagation
    of cauliflower by meristem culture. *Acta Horticulturae* 78, 289-96.
Heinz D.J., Krisnamurthi M., Nickell L.G. & Maretzki A. (1977) Cell,
    tissue, and organ culture in sugar cane improvement. *Plant Cell,
    Tissue and Organ Culture* (Ed. by J. Reinert & Y.P.S. Bajaj),
    pp.3-17. Springer-Verlag, Berlin.
Howland G.P. & Hart R.W. (1977) Radiation biology of cultured plant
    cells. *Plant Cell, Tissue and Organ Culture* (Ed. by J. Reinert &
    Y.P.S. Bajaj), pp.731-56. Springer-Verlag, Berlin.
Maliga P. (1978) Resistance mutants and their use in genetic
    manipulation. *Frontiers of Plant Tissue Culture, 1978* (Ed. by
    T.A. Thorpe), pp.381-92. International Association for Plant
    Tissue Culture, Calgary.
Murashige T. & Skoog F. (1962) A revised medium for rapid growth and
    bioassays with tobacco tissue cultures. *Physiologia Plantarum* 15,
    473-97.
Sunderland N. (1971) Anther culture: a progress report. *Science
    Progress, Oxford* 59, 527-49.
Sunderland N. & Roberts M. (1977) New approach to pollen culture.
    Nature 270, 236-8.
Weatherhead M.A. & Henshaw G.G. (1979) The production of homozygous
    diploid plants of *Solanum verrucosum* by tissue culture techniques.
    Euphytica 28, 765-8.
Westcott R.J., Grout B.W.W. & Henshaw G.G. (1979) Rapid clonal
    propagation of *Solanum curtilobum* var. Malku by aseptic shoot
    meristem culture. *The Biology and Taxonomy of the* Solanaceae (Ed.
    by J.G. Hawkes, R.N. Lester & A.D. Skelding), pp. 377-82. Academic
    Press, London.

Embryo culture in the production
of disease-resistant brassicas

C.L. ROSS
Scottish Plant Breeding Station,
Pentlandfield, Roslin, Midlothian EH25 9RF

INTRODUCTION

The cross between *Brassica campestris* (2n = 20, aa; turnips, turnip rape
and oriental salad vegetables) and *Brassica oleracea* (2n = 18, cc; the
kale and cabbage group) is made to resynthesize the naturally occurring
allotetraploid hybrid *Brassica napus* (2n = 38, aacc; swede turnip and
forage rape; U, 1935) which normally lacks some of the characters found
in the parent species. The transfer of disease resistance, especially
to *Plasmodiophora brassicae* (causing clubroot), from *B. campestris* to
the *B. napus* crops, swedes and forage rape, is one of the major reasons
for making the cross (McNaughton, 1979). *B. napus* and *B. campestris* can
be hybridized readily but it is difficult to cross *B. napus* with
B. oleracea and, as *B. oleracea* also has characters desirable in
B. napus, the direct resynthesis of *B. napus* from *B. campestris* and
B. oleracea is the best method of introgressing new and useful variation.

Unfortunately hybrid embryos, obtained from crosses between
B. campestris and *B. oleracea*, have been shown to abort at a fairly
early stage of development, due to endosperm deficiency (Håkansson,
1956). The frequency of hybrids derived from normal pollinations is,
therefore, extremely low. However, this frequency can be increased
greatly if hybrid embryos are rescued using the technique of embryo
culture.

The method described here has been slightly modified from that of
Nishi *et al.* (1961). It has been kept relatively simple, and is a
compromise between being able to rescue sufficient numbers of embryos
and avoiding both the need for complex, unstable media and the transfer
of embryos to a second medium before moving to compost. For this reason
the stage of development of the embryo at time of culture is important:
the more immature the embryo the more complex its nutrient requirements,
but at the same time there is a decrease with age in the number of
embryos remaining viable.

METHOD

Plants for cross pollination are grown in an unheated glasshouse;
controlled conditions are not required. Crosses are made between
B. oleracea and *B. campestris* at both the diploid and tetraploid levels.
(Haploid plants produced from the diploid crosses must be colchicine-
treated at a later stage to double the chromosome number.) The desired
interspecific pollinations are made, following emasculation at the bud
stage, taking the usual precautions against contamination. *B. oleracea*
is used as the female parent, the flowers being larger and easier to
emasculate and pollinate; furthermore *B. oleracea* plants live longer
and grow better under glasshouse conditions than *B. campestris* plants.

The greatest production of hybrid embryos is obtained at the first
flush of flowering, which is usually about April in the Edinburgh area.

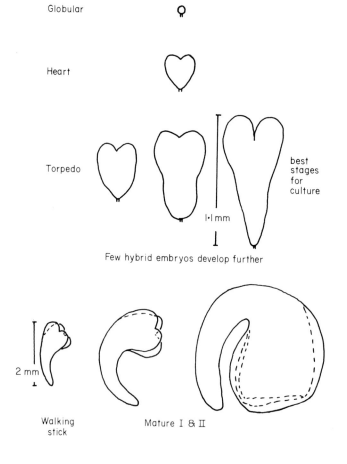

Figure 1. Stages of embryo development in *Brassica* spp.

The optimal stage of embryo development for culturing is between late
heart and late torpedo (Fig. 1). Growth conditions, particularly
temperature, have been found to affect the speed of embryo growth and
development. Capsules are therefore sampled at intervals from approximately 30 days after pollination, depending on prevailing weather
conditions. Embryos have been successfully cultured from 19 to 59 days
after pollination but the majority are obtained at the right stage at
25-45 days.

Capsules are surface sterilized by dipping in 70% ethanol and the
embryos are dissected out under aseptic conditions using a binocular
microscope, and placed on slopes of agar solidified medium (10 ml; Table
1) in 30 ml, wide-necked culture bottles. The microscope is mounted in
a perspex cover which allows access from the lower front edge only
(alternatively, a laminar flow cabinet may be used); all cleaning is
done with 95% ethanol. Instruments are sterilized by immersion in 95%
ethanol and allowed to dry without flaming.

Table 1. Medium for culture of *B. oleracea* x *B. campestris*
hybrids.

	mg/l		mg/l
KCl	850	H_3BO_3	1.5
KNO_3	800	KI	0.75
NH_4NO_3	750	$CuSO_4.5H_2O$	0.025
$MgSO_4.7H_2O$	368.5	$Na_2MoO_4.2H_2O$	0.032
$Ca(NO_3)_2.4H_2O$	287.5	Meso-inositol	100
Na_2SO_4	100	Glycine	2.0
KH_2PO_4	75	Nicotinic acid	0.5
Fe EDTA	30	Thiamine HCl	0.1
$MnSO_4.4H_2O$	6.65	Pyridoxine HCl	0.1
$ZnSO_4.7H_2O$	2.65		
Sucrose	30 g/l		
Agar	7 g/l		

The pH is adjusted to 6.0-6.5 and the medium is sterilized by
autoclaving at 121°C for 20 min.

The cultured embryos are grown either with natural daylight by a
north facing window, or under four 40 W warm white fluorescent lamps.
Plantlets obtained from embryos are left on the same slope until large
enough to transfer to UC (University of California) compost. Humidity
is reduced slowly after this transfer in a mist propagation unit.

REFERENCES

Håkansson A. (1956) Seed development of *Brassica oleracea* and *B. rapa* after certain reciprocal pollinations. *Hereditas* 42, 373-95.

McNaughton I.H. (1979) The possibilities of improved *Plasmodiophora* resistance through inter-specific and inter-generic hybridization. *Proceedings of Woronin + 100 International Conference on Clubroot, University of Wisconsin-Madison 1977*, pp. 100-112.

Nishi S., Kawata J. & Toda M. (1961) Studies on the embryo culture in vegetable crops. Part I. Embryo culture of immature crucifer embryos. *Bulletin of the National Institute of Agricultural Sciences (Japan), Series E*, 9, 58-127.

U N. (1935) Genome analysis in *Brassica* with special reference to the experimental formation of *B. napus* and peculiar mode of fertilization. *Japanese Journal of Botany* 7, 389-452.

Section 6
Concluding discussion

Challenges for the future

J.P. HELGESON

USDA, SEA, AR Plant Disease Resistance Research Unit,
Department of Plant Pathology, University of Wisconsin,
Madison, WI 53706, USA

This volume describes many tissue culture methods which are extremely useful to plant pathologists. However, it is obvious that there are substantial gaps in our knowledge of cultured cells, and that the solutions to these problems could greatly benefit plant pathologists as well as others who work with tissue cultures. In this concluding chapter I first consider four such problem areas, and finally mention the potential value of pathological systems in solving basic problems in other areas of plant biology.

Biochemistry and physiology of cultured cells and protoplasts
Our knowledge of biochemical pathways in cultured cells is incomplete. We do not know, for example, the details of synthetic pathways of some phytoalexins (Dixon, this volume) nor do we know the factors that control such pathways. Also, we have not yet identified the molecules involved in the dialogue between host and pathogen. In some instances events in cultures seem to correlate well with those in intact plants (Helgeson & Haberlach, this volume) whereas in others such correlations do not exist (see the Introduction by Ingram, this volume). Tissue cultures can provide excellent material for biochemical studies but progress in this area is slow.

Stress physiology   Survival of cells at low temperatures can be a problem (Henshaw *et al.*; Withers; this volume). Pollen in culture can be very sensitive to stress (Sunderland, this volume). These are only two facets of a general problem; the whole subject of the effects of stress on cell viability and on differentiation is in great need of quantitative work. Protoplasts, for example, are subjected to prolonged water stress from the various osmotica used. With leaf tissue *in vivo* such stress would cause photosynthesis and protein synthesis to stop, abscisic acid to build up, polyribosomes to disperse, nitrate reductase to cease functioning and many other malfunctions to occur (Boyer, 1973; Hsaio, 1973). It is not surprising, therefore, that many protoplasts either do not survive in culture or lose the capability for differentiation. The physiology and biochemistry of stress would seem to be a

Ingram D.S. & Helgeson J.P. (1980) *Tissue Culture Methods for Plant Pathologists*

fruitful area for research and tissue culture systems would seem to be ideal for such studies.

Differentiation of cells   The problem of cell differentiation is a particularly critical one. Many species can be regenerated from tissue cultures but only a few, as yet, will yield plants from protoplasts (Thomas *et al.*, this volume). In 1957 Skoog & Miller published their account of the effects of auxin/cytokinin ratios on the differentiation of tobacco tissue cultures, and their work has guided much of the work on differentiation since that time. A number of techniques have been devised to test the effects of various balances of auxin and cytokinin on cell morphology (see, for example, Potrykus *et al.*, 1978), but it is probably the internal concentration of phytohormones, as affected both by external conditions and internal controls, which determines the course of differentiation. The challenge here is to determine these internal balances, concentrations and even types of phytohormones, and to ascertain which changes actually lead to differentiation. The chemical methods are now available. High performance liquid chromatography and gas-liquid chromatography have been used with the major phytohormones and affinity columns have been devised to aid pre-purification (Constantinidou *et al.*, 1978). Perhaps it is possible to "push" the recalcitrant species towards differentiation once the effect of external treatments on internal balances is known.

Genetics of cultured cells   Genetic stability is a major problem. In many cases the chromosome number of cells may change after periods in culture, and such changes are often correlated with the loss of capability for differentiation (Murashige & Nakano, 1967). Selections for drug or antimetabolite resistance in cultured cells can lead to temporary changes in biochemical pathways rather than stable genetic changes. Similarly the crown-gall system might be useful for the genetic engineering of plant cells but, as yet, it appears that the information imparted by the bacterial plasmid does not survive meiosis (Davey *et al.*; Day; this volume). At times there appears to be substantial diversity in protoplasts such as those from potato (Shepard *et al.*, 1980), but at other times such differences in protoplasts from the same species are not apparent (Wenzel *et al.*, 1979). It is clear from these examples that much still remains to be done before the genetics of cultured cells are fully understood.

The pathological connection   This book stresses the usefulness of tissue cultures to plant pathologists. It might be appropriate, however, to end on a different note. Pathological systems have been and could continue to be highly useful to biochemists, physiologists and others working with cell cultures. Toxins from various disease-inducing organisms can provide effective screens for changed cells in a population. Also, if one is to improve cells via tissue culture techniques the addition of a disease resistance factor would be a major advantage. Finally, plant pathogens may be the only organisms that can bypass a cell's defence against foreign materials; hence modified pathogens might be vehicles for crop improvement, and physiological responses involved in resistance or susceptibility might be crucial for success. Thus in many areas the plant pathologist can both contribute to and benefit from the development of tissue culture techniques.

REFERENCES

Boyer J.S. (1973) Response of metabolism to low water potentials in plants. *Phytopathology* 63, 466-72.
Constantinidou H.A., Steele J., Kozlowski T.T. & Upper C.D. (1978) Binding specificity and possible analytical applications of the cytokinin-binding antibody, anti N^6-Benzyl adenosine. *Plant Physiology* 62, 968-74.
Hsaio T.C. (1973) Plant responses to water stress. *Annual Review of Plant Physiology* 24, 519-70.
Murashige T. & Nakano R. (1967) Chromosome complement as a determinant of the morphogenetic potential of tobacco cells. *American Journal of Botany* 54, 963-70.
Potrykus I., Harms C. & Lorz H. (1978) Multiple-drop-array (MDA) technique for the large scale testing of culture media variations in hanging microdrop cultures of single cell systems. I. The technique. *Plant Science Letters* 14, 231-5.
Shepard J., Bidney D. & Shahin E. (1980) Potato protoplasts in crop improvement. *Science* 208, 17-24.
Skoog F. & Miller C. (1957) Chemical regulation of growth and organ formation in plant tissues cultured *in vitro*. *Symposia of the Society for Experimental Biology* 11, 118-31.
Wenzel G., Schieder O., Przewozny T., Sopory S. & Melchers G. (1979) Comparison of single cell culture derived *Solanum tuberosum* L. plants and a model for their application in breeding programs. *Theoretical and Applied Genetics* 55, 49-55.

Index